KV-579-737

Multiple Choice
Questions in Physiology

Multiple Choice
Questions in Physiology

With answers and explanatory comments

Lynn Bindman
Peter Ellaway
Brian Jewell
Laurence Smaje

The Department of Physiology,
University College London

Edward Arnold
A division of Hodder & Stoughton
LONDON BOSTON MELBOURNE AUCKLAND

© 1993 Lynn Bindman, Peter Ellaway, Brian Jewell & Laurence Smaje

First published in Great Britain 1993

Distributed in the Americas by Little, Brown and Company
34 Beacon Street, Boston, MA 02108

British Library Cataloguing in Publication Data

Bindman, Lynn
  Multiple Choice Questions in Physiology.
  — 2Rev. ed
  I. Title
  574.1

  ISBN 0-340-57926-9

Whilst the advice and information in this book is believed to be
true and accurate at the date of going to press, neither the
authors nor the publisher can accept any legal responsibility or
liability for any errors or omissions that may be made.

Typeset in 8/9 pt Helvetica Light by Colset Pte Ltd, Singapore
Printed and bound in Great Britain for Edward Arnold, a division
of Hodder and Stoughton Limited, Mill Road, Dunton Green,
Sevenoaks, Kent TN13 2YA by Biddles Limited, Guildford and
King's Lynn.

Contents

Abbreviations

Multiple Choice
Questions in Physiology

Preface

Aims. Multiple-choice questions are widely used in examinations at all levels, but they often test little more than factual recall. The main purpose of this booklet is to provide a selection of questions that also test reasoning power and ability to interpret data or perform simple calculations. Most of the questions were devised originally for examining preclinical medical students at University College London and they have been found to discriminate well among students of differing abilities as judged by other criteria.

We hope that teachers of physiology and their students will benefit from this booklet. Teachers may find the questions provide a useful source of ideas for creating their own questions, while students will gain experience of multiple-choice questions that test understanding as well as factual knowledge. Students may also find working through the booklet a useful form of revision as explanatory comments are provided for most of the answers; these are more detailed for points that students tend to find difficult.

How to use the booklet. The main topics covered are set out on the Contents page. Earlier editions of this book did not include questions on the central nervous system and special senses because the topics were covered in a separate MCQ book on Neurophysiology. This new, expanded edition includes questions on all important aspects of neurophysiology taken from that book, and hence it covers the entire preclinical physiology course. Questions on applied and clinical physiology that are more suitable for second year medical students are marked with an asterisk. Every question consists of a stem and five statements, each of which must be judged '**True**' or '**False**'. We suggest you answer all parts of the question before looking at the answers and explanations given on the opposite page.

Test yourself. You may like to compare your performance with that of our medical students. Give yourself a mark of +1 for each correct judgement, −1 for each incorrect judgement, and zero for any you left out. The score for each question can therefore vary from +5 to −5 marks. In tests consisting of 20 questions, the mean mark for University College students in examinations over a period of five years has been around +50 with a standard deviation of about 10 marks.

Ambiguous Questions. It is extremely difficult to set questions that contain no ambiguities and for several years we have invited our examination candidates to qualify their answers to questions they find ambiguous by writing on the question paper. In fact this has necessitated very few adjustments to the marks awarded, but it has allowed us to refine some of our questions and it has undoubtedly helped to reduce the tension often induced by multiple-choice examinations. We would be pleased if users of this booklet would draw our attention to ambiguities in the questions included in it.

Acknowledgements. We thank the many colleagues and students at University College London and elsewhere who have criticised earlier versions of our text. They are not responsible for any errors that have crept in subsequently. We are grateful to Professor E.F. Evans for permission to use his results in the figure of question 117. Finally we thank Audrey Besterman for her skilful art work.

1 The following statements give reasons for including certain ingredients in physiological bathing solutions:
   (a) The composition of a physiological bathing solution should be as close as possible to that of intracellular fluid.
   (b) Nerve axons become inexcitable if they are bathed in a Na^+-free Ringer's solution.
   (c) Spontaneous activity will occur in nerve fibres if the Ca^{2+} concentration in the bathing medium is too high.
   (d) The membrane potential of a nerve axon is more sensitive to the concentration of potassium in the bathing medium than to the concentration of any other ion.
   (e) If a bathing solution containing bicarbonate is bubbled with pure oxygen instead of 5% CO_2/95% O_2 mixture, its pH will be too low.

2 The following are statements about the osmotic pressure of plasma:
   (a) The *total* osmotic pressure of plasma is similar to that of 0·9% NaCl.
   (b) The *total* osmotic pressure of plasma is similar to that of 0·9% glucose.
   (c) The *total* osmotic pressure of the plasma is largely due to the contributions of $Na^+ + Cl^-$ ions.
   (d) The *colloid* osmotic pressure of plasma is about 25 mm Hg.
   (e) The *total* osmotic pressure of plasma opposes the ultrafiltration of fluid from the capillaries.

3 Glycerol penetrates the red cell membrane rather slowly. Which of the following will happen when red cells are suspended in a solution of 1 mol/l glycerol in water?
   (a) The red cells will immediately undergo haemolysis.
   (b) The red cells will shrink, becoming permanently crenated.
   (c) The red cells will swell first and then shrink to become permanently crenated.
   (d) The red cells will shrink first and then swell and haemolyse.
   (e) No volume changes will take place.

4 The following are statements about human red blood cells:
   (a) Red cells are rigid biconcave discs.
   (b) Normally 10 to 20% of circulating red cells contain remnants of nuclear material.
   (c) Following haemolysis, red cells release haemopoietin which stimulates the production of more red cells.
   (d) Red cells contain carbonic anhydrase.
   (e) Red cells make a major contribution to the buffering capacity of the blood.

1 (a) **False** It should resemble interstitial fluid.
  (b) **True** Na is essential for the action potential.
  (c) **False** Spontaneous activity will occur if $[Ca^{2+}]$ is too *low*. See the answer to 96 (c).
  (d) **True** This is because the resting potential depends mainly on the ratio of the K^+ concentrations inside and outside the membrane.
  (e) **False** Its pH will be too high.

$$pH = pK + \log_{10} \frac{[HCO_3^-]}{[CO_2]}$$

2 (a) **True** Red blood cells will neither swell nor shrink when placed in $0 \cdot 9\%$ NaCl which is often referred to as 'physiological' saline. Both have an osmolarity of about 300 mosmol/l.
  (b) **False** The molecular weight of glucose is different from that of NaCl, so $0 \cdot 9\%$ glucose will have a different molarity. Even if the molarities of the two solutions were equal, the osmolarities would differ because NaCl dissociates in solution and exerts a higher osmotic pressure.
  (c) **True**
  (d) **True** The plasma proteins exert a very small fraction (about 1/200) of the total osmotic pressure of about 5000 mm Hg (1 mm Hg $\equiv$ 133$\cdot$3 Pa).
  (e) **False** It is only the colloid osmotic pressure that opposes the loss of fluid. Small particles like ions and glucose can diffuse across the capillaries and hence can exert no transmural osmotic pressure.

3 (a) **False** There are neither chemical nor physical factors that will cause *immediate* disruption of the membrane.
  (b) **False**
  (c) **False**
  (d) **True** The osmotic gradient between the glycerol solution (1 osmol/l) and the cell contents ($0 \cdot 3$ osmol/l) will initially cause water to leave the red blood cell. Glycerol penetrates slowly under the concentration gradient until the osmolarity of the cell contents exceeds that of the glycerol. Water will therefore re-enter the cell and eventually the swelling produced by the continual entry of glycerol and water will burst the cell membrane.
  (e) **False**

4 (a) **False** They are not rigid, and indeed undergo considerable reversible deformation as they pass through capillaries. They only appear as biconcave discs when unstressed.
  (b) **False** Normally there are 1 to 2% of reticulocytes. Much higher percentages occur when there is accelerated haemopoiesis.
  (c) **False** The main source of haemopoietin is the kidney.
  (d) **True**
  (e) **True**

*5 A blood count in a woman aged 40 gave the following picture: Hb, 110 g/l, RBC, $3\cdot0 \times 10^{12}$/l; mean cell diameter, $8\cdot2\mu$m. The following are statements about the findings:
  (a) The blood picture is within normal limits.
  (b) The findings are typical of iron deficiency anaemia.
  (c) The findings are typical of vitamin B_{12} deficiency.
  (d) This blood would carry about 150 ml oxygen/l blood.
  (e) The findings are typical of someone living at high altitude.

*6 Oxygen delivery to the tissues is usually reduced in the following conditions:
  (a) Sickle cell anaemia.
  (b) Reduced ventilation/perfusion ratio.
  (c) Severe iron deficiency anaemia.
  (d) Congestive cardiac failure.
  (e) Emphysema.

7 The graph shows the relationship between the saturation of haemoglobin with oxygen and the partial pressure of oxygen.

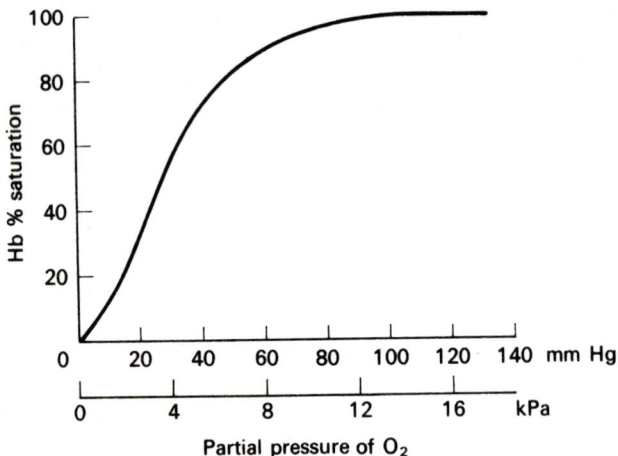

  (a) A rise in pH will move the curve to the left.
  (b) Anaemia will depress the curve.
  (c) A fall in temperature will move the curve to the left.
  (d) An increase in 2, 3, diphosphoglycerate in the red cells would shift the curve to the right.
  (e) Foetal haemoglobin has a similar dissociation curve.

5 (a) **False** The Hb is lower, the RBC count lower and mean cell diameter larger than normal.
  (b) **False** The woman is anaemic, but in iron deficiency anaemia red blood cells are smaller than usual.
  (c) **True** Vitamin B_{12} deficiency produces a megaloblastic anaemia which results in RBCs that are larger than normal.
  (d) **True** Each 1 g of haemoglobin can carry 1·34 ml of oxygen at NTP.
  (e) **False** In acclimatization to high altitude the red cell count and haemoglobin concentration would be higher than normal.

6 (a) **True** Sickled cells are more rigid than normal cells, and cannot therefore squeeze their way through narrow capillaries. (The abnormal Hb-S molecules form polymers when the oxygen tension is low, and the cell becomes sickle-shaped).
  (b) **True** Blood leaving the lungs is likely to be less well oxygenated when the ventilation/perfusion ratio falls.
  (c) **True**
  (d) **True** The cardiac output is reduced and the peripheral resistance vessels are likely to be constricted. Both factors reduce oxygen delivery to the tissues.
  (e) **True** As in (b) the blood is likely to be incompletely oxygenated when it leaves the lungs.

7 (a) **True**
  (b) **False** Look again at the ordinate! It is *percentage* saturation of haemoglobin, not oxygen content of the blood.
  (c) **True**
  (d) **True** This occurs for example in acclimatization and in anaemia, and assists in unloading oxygen in the tissues.
  (e) **False** It is displaced to the left, thereby facilitating oxygen transfer from mother to foetus. Foetal Hb is less affected by 2,3.DPG than adult Hb.

8 This diagram shows how the oxygen content of a sample of blood varied with the PO_2 of the gas with which it was in equilibrium at a PCO_2 of 40 mm Hg (5·3 kPa).

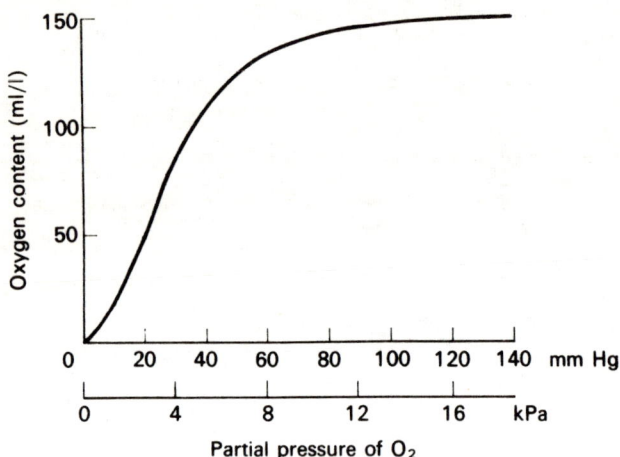

(a) The sigmoid shape of the curve is due to the fact that red blood cells vary in their affinity for oxygen.
(b) This sample of blood must have had a haemoglobin concentration of about 150 g/l.
(c) The haemoglobin is almost fully saturated at a PO_2 of 100 mm Hg (13·3 kPa).
(d) If the PO_2 had been raised to higher values, the curve would have become horizontal when the PO_2 exceeded 150 mm Hg (20 kPa).
(e) If the PCO_2 of the sample is increased, the oxygen content would be greater at a given PO_2.

9 The following are statements about the viscosity of blood:
(a) Viscosity of the blood increases as the haematocrit increases.
(b) The viscosity of blood flowing through small tubes (e g. arterioles) is lower than in large tubes (e.g. the aorta).
(c) The viscosity is less at 25°C than at 37°C.
(d) Viscosity of the blood usually increases in acclimatised mountaineers.
(e) It is decreased in people with iron-deficiency.

10 Blood clotting is delayed or prevented *in vitro* when:
(a) Blood is placed in polythene tubes (compared with glass tubes).
(b) The temperature of the blood is increased from room temperature to 37°C.
(c) Sodium citrate is added to the blood.
(d) Dicoumarol is added to the blood.
(e) When heparin is added to the blood.

8 (a) **False** It is due to the binding properties of the haemoglobin molecule for oxygen.
  (b) **False** 1 g haemoglobin can combine with 1·34 ml of oxygen at NTP.
  (c) **True**
  (d) **False** The curve (i.e. the oxygen content) would continue to rise slightly as more oxygen is dissolved in physical solution.
  (e) **False** It's the other way around (Bohr effect).

9 (a) **True** Remember that the haematocrit is the fraction of red cells in the blood sample.
  (b) **True** A surprising fact with a complicated explanation.
  (c) **False** Viscosity always increases with a fall in temperature.
  (d) **True** In acclimatization the haematocrit rises.
  (e) **True** Iron-deficiency leads to anaemia and hence to a lower haematocrit.

10 (a) **True** The non-wettable surface delays surface activation of the initial sequences of blood clotting.
  (b) **False** The optimum temperature for the enzymes involved is around 37°C.
  (c) **True** It binds the calcium ions that are essential for clotting.
  (d) **False** Dicoumarol is a competitive antagonist of vitamin K which is required for prothrombin synthesis. Dicoumarol added to the blood *in vitro* therefore has no effect on blood clotting as prothrombin is already present. *In vivo* it needs to be taken for a few days before any effects are seen.
  (e) **True** Heparin is an effective anticoagulant both *in vitro* and *in vivo* and probably acts as an anti-thrombin.

*11 It is often difficult to find suitable blood for transfusion to patients who have had:
  (a) A previous transfusion of Rh+ blood.
  (b) Horse serum injections (e.g. anti-tetanus serum).
  (c) Many previous transfusions.
  (d) Syphilis or jaundice.
  (e) No previous transfusion.

*12 If Rh+ blood is transfused into an Rh− woman who has not previously been transfused, then:
  (a) Anti-Rh antibodies will be produced by the woman.
  (b) The bloods are incompatible so red cell agglutination and death may follow.
  (c) In a subsequent pregnancy the foetus could be threatened by haemolytic disease.
  (d) There is no immediate or long-term effect as 70% of the Rh+ population are heterozygous.
  (e) Provided anti-D antibody is given before the next pregnancy no harm will be done.

13 A man of blood group A has 2 children. Plasma from the blood of one of them agglutinates his red cells while that from the other does not.
  (a) Father *must* be heterozygous group A.
  (b) Children *must* have had different mothers.
  (c) 'Agglutinating' child *could* be group O.
  (d) Mother of 'agglutinating' child *must* be group O.
  (e) 'Non-agglutinating' child *could* be group AB.

14 The following are statements about the autonomic nerve supply to the heart:
  (a) In the normal animal there is a background level of sympathetic 'tone' to the heart.
  (b) Increased sympathetic activity decreases the rate of firing of the pacemaker cells.
  (c) Sympathetic stimulation shifts the curve relating stroke work (ordinate) to diastolic volume of the heart (abscissa) to the right.
  (d) Sympathetic simulation increases the rate of coronary blood flow.
  (e) Stimulation of the vagus nerve slows the heart rate.

11 (a) **False** If the recipient were Rh + ve, a Rh + ve transfusion would not provoke antibody formation. If the recipient were Rh − ve, a previous Rh + ve transfusion would have provoked antibody production but Rh − ve blood is readily available, so there is no problem.

(b) **False** This could make subsequent injections of horse serum dangerous but has no effect on blood group antigen/antibody reactions.

(c) **True** Each transfusion increases the possibility of antibody formation because of subgroup incompatibility.

(d) **False** People who have had syphilis or jaundice should not be used as *donors.*

(e) **False** But in every case recipient and donor blood should be directly cross-matched even if apparently compatible.

12 (a) **True** These will be mostly anti-D.

(b) **False** There are no naturally occurring anti-Rh antibodies.

(c) **True** After the transfusion anti-Rh + antibodies would be formed; they may cross the placental barrier and if the foetus were Rh + ve react with foetal red blood cells.

(d) **False** Irrelevant.

(e) **False** Anti-D antibody cannot be used in this situation. It can be used immediately after childbirth to prevent a Rh − ve mother from forming anti-D antibodies to cells from a Rh + ve foetus that could have passed into her circulation at parturition. The donated antibodies are destroyed within a few weeks and so will not threaten future foetuses.

13 (a) **True** The A gene is dominant, so if he were *homozygous* group A, his children would be group A or AB, and neither plasma would agglutinate his cells.

(b) **False** The mother could for example be group AB. The non-agglutinating child could be group AA and the agglutinating child, BO.

(c) **True** Or the child could be group B with B gene inherited from the mother and O from the father.

(d) **False** See answer to (b).

(e) **True**

14 (a) **True**

(b) **False** It increases heart rate.

(c) **False** The curve is shifted upwards and to the left (greater stroke work at given fibre length).

(d) **True**

(e) **True**

15 An infusion of noradrenaline is given to a human subject at a sufficient rate to produce a rise in systolic BP of 20 mmHg (2·7 kPa). The consequences are likely to be:
   (a) An increase in diastolic blood pressure.
   (b) Decreased firing in the baroreceptor nerves.
   (c) A reflex bradycardia (slowing of the heart).
   (d) A generalized decrease in sympathetic nerve discharge.
   (e) A decrease in [FFA] (free fatty acid concentration).

16 Cardiac output is decreased:
   (a) During stimulation of sympathetic nerves to the heart.
   (b) As a consequence of decreased pressure in the carotid sinus.
   (c) By increasing the end-diastolic volume of the heart.
   (d) On cutting the vagal nerves to the heart.
   (e) On standing up.

17 The consequences of arteriolar vasoconstriction in an organ are likely to be:
   (a) A reduction in blood flow through the organ.
   (b) An increase in capillary pressure in the vascular bed.
   (c) A decrease in the arterio-venous oxygen difference (i.e. difference in oxygen concentration in blood entering and leaving the organ).
   (d) An increase in the partial pressure of CO_2 in blood leaving the organ.
   (e) A decrease in the rate of lymph flow from the organ.

18 The following are statements about the exchange of substances between blood and the interstitial fluid:
   (a) O_2 and CO_2 pass easily across capillary walls.
   (b) The exchange of non lipid-soluble substances across capillary walls requires the presence of water-filled pores in the endothelial lining.
   (c) Movement of a substance from blood to interstitial fluid will occur only if there is a favourable concentration gradient.
   (d) Exchange diffusion provides an enormously greater turnover of water between the blood and the interstitial space than filtration and reabsorption along the lines envisaged by Starling.
   (e) Exchange of substances between blood and interstitial fluid also occurs across the walls of venules.

15 (a) **True**
   (b) **False** An *increased* firing results from the rise in blood pressure.
   (c) **True** A consequence of the increased baroreceptor discharge; vagal slowing of the heart overrides any direct cardioaccelerator action of noradrenaline.
   (d) **True** This is another reflex consequence of (b).
   (e) **False** A *rise* in |FFA| is brought about by activation of β receptors. Noradrenaline acts mainly on α receptors.

16 (a) **False** Heart rate and stroke volume rise. hence cardiac output increases.
   (b) **False** There is a reflex increase in cardiac output (and peripheral resistance).
   (c) **False** It is increased by the Starling relationship except in the failing heart.
   (d) **False** The heart rate increases because of the abolition of vagal tone. hence cardiac output increases.
   (e) **True** Cardiac output drops when you stand up due to pooling of blood and remains lower in spite of compensatory reflex adjustments.

17 (a) **True** Flow is given by P/R. The vascular resistance of the organ. R. is increased by arteriolar constriction. but P. the driving pressure. will not be increased much by vasoconstriction in a single organ. Flow will therefore decrease.
   (b) **False**
   (c) **False** If the blood flow falls. *more* oxygen will be extracted during the passage of blood through the capillary.
   (d) **True**
   (e) **True** There will be diminished filtration if capillary transmural pressure is reduced.

18 (a) **True** These gases are lipid soluble and can therefore easily penetrate cell membranes.
   (b) **True** However such pores do not have to have a permanent existence.
   (c) **False** Diffusion occurs in both directions but *net* movement by diffusion can only occur *down* a concentration gradient. Note also that movement can take place *against* a concentration gradient by ultrafiltration or by active transport.
   (d) **True**
   (e) **True** Total surface area of venules approaches that of capillaries and filtration coefficients can be higher.

19 The following are important variables in circulatory physiology: cardiac output, CO; total peripheral resistance, TPR; mean arterial blood pressure, BP; stroke volume, SV; heart rate, HR. Are the following relationships true or not?
(a) BP = CO × TPR
(b) CO = BP/TPR
(c) CO = SV × HR
(d) HR = BP/(SV × TPR)
(e) TPR = BP × SV × HR

20 The diagram shows a tube in which there is a narrow region, B, where the diameter is half that at A and C.

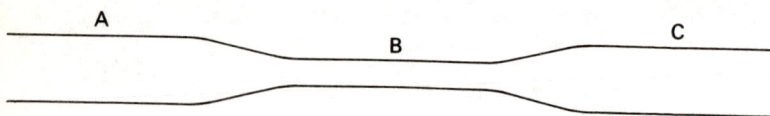

(a) The resistance to flow (per unit length of tube) will be eight times greater at B than at A.
(b) The velocity of flow will be four times greater at B than at A.
(c) The lateral (side-wall) pressure at B will be greater that at A.
(d) The pressure drop along the narrow section will be inversely proportional to its length.
(e) The tangential stress in the wall of the tube will be less at B than at A.

21 The following are statements about veins:
(a) Kinetic energy accounts for a much higher proportion of the total energy of blood flowing in the inferior vena cava than in does in the aorta.
(b) Sub-atmospheric pressures are never found in blood vessels outside the thoracic cavity.
(c) If the transmural pressure at a point A in a blood vessel is greater than that at point B, then the direction of blood flow in the vessel must be from A to B.
(d) The pressure in the veins of the foot will be lower when a person is walking than it will be when he is standing still.
(e) About two thirds of all the blood in the body is found in the systemic veins.

19 (a) **True** This can be considered as a haemodynamic equivalent of Ohm's law.
   (b) **True**
   (c) **True**
   (d) **True**
   (e) **False**

20 (a) **False** Resistance is proportional to 1/radius^4, so if the radius is halved resistance will increase by a factor of 16.
   (b) **True** Velocity × cross-sectional area must be the same at A and B, and the area decreases by a factor of 4.
   (c) **False** If velocity (v) is greater at B than at A, the kinetic energy of the moving fluid ($\frac{1}{2}$ mv^2) will also be greater. As the total energy of fluid (kinetic + pressure energy) cannot be greater at B than at A, pressure energy at B (and therefore side wall pressure) must be less than at A.
   (d) **False** The pressure difference will be proportional to length (Poiseuille's Law).
   (e) **True** The Laplace relation predicts this.

21 (a) **True** The flow (ml/min) in both vessels is about the same, so is their cross-sectional area; therefore mean velocity and kinetic energy will be similar. However pressure energy is nearly 100 times greater in the aorta than in the inferior vena cava.
   (b) **False** They occur in cranial sinuses, where rigid walls prevent vessels collapsing due to sub-atmospheric pressure.
   (c) **False** *Transmural* pressure gradients have nothing to do with blood flow *along* tubes.
   (d) **True** This is the result of muscle pumping and the presence of valves.
   (e) **True**

22 A and B represent two vascular beds (e.g. skin and muscle of upper limb) which
    are supplied with blood from the same artery (e.g. brachial). F denotes blood flow
    and R resistance to blood flow. Suppose the pressure drop, ΔP, across the
    vascular beds remains constant:

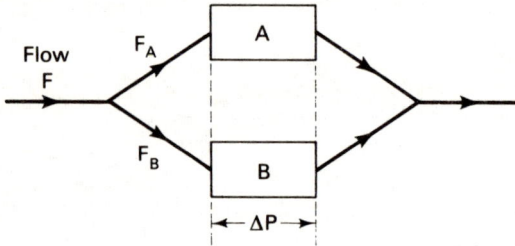

(a) The flow through A is given by: $F_A = \Delta P/R_A$
(b) The total flow is given by: $F = F_A + F_B = \Delta P/(R_A + R_B)$
(c) An increase in R_B will lead to a reduction in F_B.
(d) An increase in R_B will lead to an increase in F_A.
(e) The combined resistance, R, of the two vascular beds is given by: $R = 1/R_A + 1/R_B$.

23 Diagrams a and b are records obtained by venous occlusion plethysmography
    on the human forearm taken before and during (or just after) an experimental
    intervention:

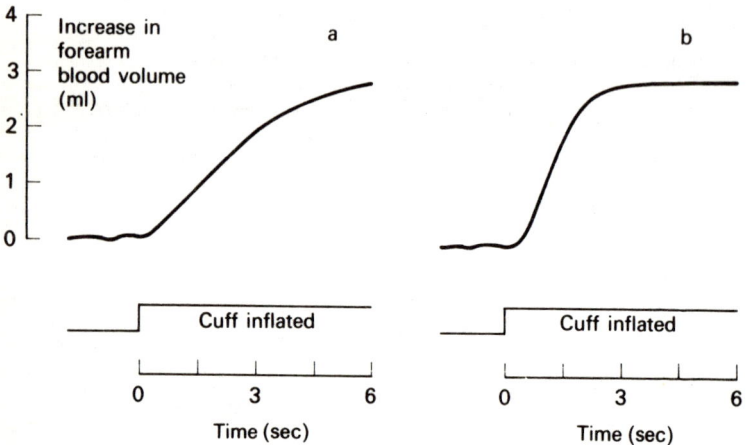

(a) Recording b could have been taken after cooling the forearm.
(b) Recording b could have been taken during fainting.
(c) Recording b could have been taken immediately after exercising the forearm.
(d) Recording b could have been taken after administration of a β-adrenergic
     antagonist.
(e) Recording b could have been taken by raising the occlusion cuff pressure to
     60 mm Hg instead of the 40 mm Hg used in the control record a (to 8 kPa
     from 5.3 kPa).

22 (a) **True** The plumbing version of Ohm's Law again.

    (b) **False**
$$F = \frac{\Delta P}{R_A} + \frac{\Delta P}{R_B} \neq \frac{\Delta P}{R_A + R_B}$$

    (c) **True** Follows from relationship described in (a).

    (d) **False** If ΔP remains constant. F_A will not be affected by R_B. However the total flow. F. will be reduced.

    (e) **False** True statements are:
$$\frac{1}{R} = \frac{1}{R_A} + \frac{1}{R_B} \qquad R = \frac{R_A R_B}{R_A + R_B}$$

23 (a) **False** The increased slope in b results from an increased forearm blood flow (cooling would decrease flow).

    (b) **True** Fainting due to a vaso-vagal attack is accompanied by 'active' vasodilation in the vascular beds of skeletal muscle.

    (c) **True**

    (d) **False** An increased blood flow could result from abolition of vasoconstrictor tone. but that would require an α-adrenergic antagonist.

    (e) **False** Changing the cuff pressure alters the final level reached on the records. but not the initial slope.

24 The left ventricle has a thicker wall than the right because:
   (a) It has to eject blood against a higher pressure.
   (b) It has to eject a greater stroke volume.
   (c) It has to do more stroke work.
   (d) It has to eject blood through a narrower orifice.
   (e) It has to eject blood at a much higher velocity.

25 This diagram shows how the pressure and volume of a ventricle change with respect to one another during the cardiac cycle.

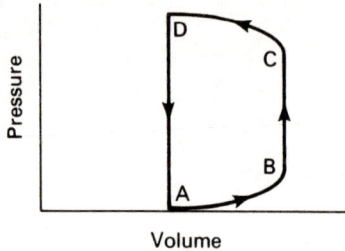

Volume

   (a) The curve between A and B is part of the passive pressure–volume relation of ventricle.
   (b) Between B and C the ventricle fills with blood.
   (c) Ejection of blood from the ventricle occurs between C and D.
   (d) The area of the loop gives a good indication of the work done by the ventricle during one cardiac cycle.
   (e) The cardiac output can be obtained by multiplying the area of the loop by the heart rate.

26 The graphs show the relation between the stroke work of the left ventricle and the pressure in the ventricle at the end of diastole.

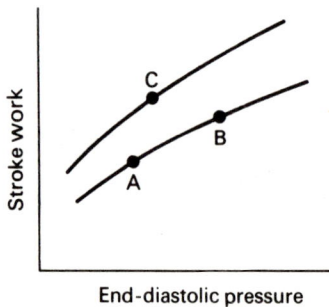

End-diastolic pressure

   (a) The end-diastolic pressure gives an indication of the size of the ventricle at the beginning of systole.
   (b) A change in stroke work from A to B could be explained by Starling's Law of the Heart.
   (c) The two graphs show different contractile states of the ventricle.
   (d) A change in stroke work from B to C could be produced by stimulating the sympathetic nerve supply to the heart.
   (e) A change in stroke work from A to C could be produced by a change in posture from lying to standing.

24 (a) **True** Remember that pressures in the pulmonary artery are about 1/6 of those in the aorta. A higher pressure requires greater (tangential) wall stress and more muscle fibres in parallel are needed to produce this.

(b) **False** Stroke volumes of R and L ventricles must be identical in the long run.

(c) **True** Main component of stroke work = stroke volume × mean pressure in outlet vessel during ejection.

(d) **False**

(e) **False** Consider answers to (b) and (d): If the same stroke volume is ejected through the same sized orifice. the mean velocity will be the same for both ventricles.

25 (a) **True** Pressure rises as relaxed ventricle fills with blood.

(b) **False** Ventricle cannot be filling if no volume change occurs: this is the phase of isovolumic contraction.

(c) **True** Note that the pressure increases during ejection.

(d) **True** Area gives pressure work done during cardiac cycle (see 24(c)). Note that the ventricle also imparts kinetic energy to blood ejected.

(e) **False** Area × heart rate = power output (i.e. work output per unit time).

26 (a) **True**

(b) **True**

(c) **True**

(d) **True**

(e) **False** While there is an increase in sympathetic tone on standing. there is a fall in end-diastolic volume (and therefore in end-diastolic pressure).

27 These diagrams show the effect of certain procedures on the blood flow to
   muscle.

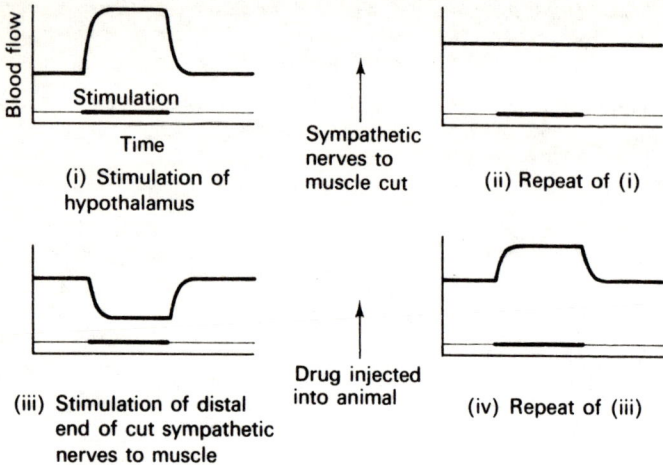

(i) Stimulation of
   hypothalamus

Sympathetic
nerves to
muscle cut

(ii) Repeat of (i)

(iii) Stimulation of distal
     end of cut sympathetic
     nerves to muscle

Drug injected
into animal

(iv) Repeat of (iii)

(a) There was vasoconstrictor tone in the muscle before the nerves were cut.
(b) The sympathetic nerve supply to the muscle contained both vasoconstrictor
    and vasodilator fibres.
(c) Activation of sympathetic vasodilator fibres is possible by stimulation of the
    hypothalamus.
(d) The drug injected between (iii) and (iv) is likely to have been a β-adrenergic
    antagonist.
(e) The change in blood flow produced by stimulation of the hypothalamus could
    have been due to the systemic release of a vasoactive substance.

28 This is a schematic diagram of the relation between blood flow through a vascular
   bed and the pressure drop across it (ΔP).

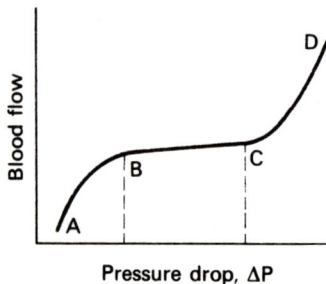

Pressure drop, ΔP

(a) This graph shows that the flow is proportional to the pressure drop across
    the vascular bed.
(b) This pressure–flow relation shows that autoregulation occurs in the vascular
    bed.
(c) This pressure–flow relation is typical of that seen in the pulmonary vascular
    bed.
(d) Within the region B–C, the resistance to flow increases as ΔP increases.
(e) A pressure–flow relation of the form shown is only obtainable if the autonomic
    nerve supply to the vascular bed is intact.

27 (a) **True** Compare the basal flows in (ii) and (i).
   (b) **True** Absence of effect of stimulation in (ii) shows that the sympathetic nerves included vasodilator fibres: the occurrence of effect (iii) shows that sympathetic nerves included vasoconstrictor fibres.
   (c) **True** See (i) and (ii).
   (d) **False** The drug has prevented vasoconstriction and is therefore likely to have been an α-adrenergic antagonist.
   (e) **False** Procedure (ii) showed that the cause of the vasodilatation in (i) was neural not humoral.

28 (a) **False** A proportional relationship must be linear: in fact the curve is S-shaped.
   (b) **True** i.e. blood flow varies very little with pressure over the range B–C.
   (c) **False** In the pulmonary vascular bed blood flow increases with very little change in pressure over the physiological range. In the diagram opposite it is *flow* that remains fairly constant.
   (d) **True** (Can you suggest a possible mechanism?).
   (e) **False** It can be found in the isolated artificially-perfused kidney.

29 This is a schematic diagram of a typical electrocardiogram from lead II for a single
   heart beat.

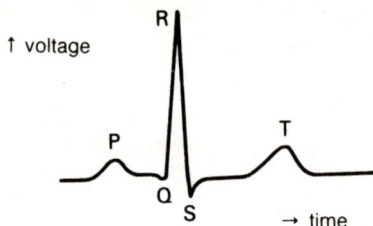

(a) The P wave is produced by depolarisation of the atria.
(b) The QRS complex is produced by depolarisation of the ventricles.
(c) The Q-T interval gives a rough indication of the duration of ventricular systole.
(d) The first heart sound occurs at about the same time as the P wave.
(e) The second heart sound occurs at about the same time as the QRS complex.

*30 Study the accompanying electrocardiograms and consider whether the following
   statements are true.

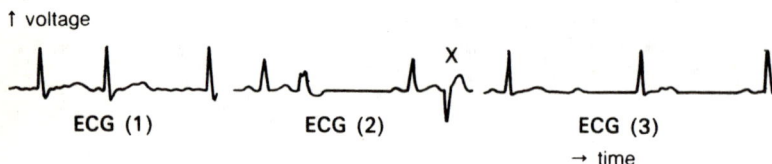

(a) ECG (1) shows sinus arrhythmia.
(b) ECG (2) shows atrial fibrillation.
(c) X in ECG (2) is a ventricular extrasystole.
(d) ECG (3) shows complete heart block.
(e) The patient from whom ECG (3) was taken would be unable to increase his
    heart rate substantially on exercise.

31 Which of the following are necessary to calculate cardiac output using the direct
   Fick method based on oxygen measurements?
   (a) O_2 consumption per minute.
   (b) Arterial O_2 concentration.
   (c) Ventilation/perfusion ratio.
   (d) Mixed venous O_2 concentration.
   (e) Diffusing capacity for O_2.

29 (a) **True**
   (b) **True**
   (c) **True**
   (d) **False** The first sound occurs as the ventricles contract and is mainly due to closure of the A-V valves.
   (e) **False** The *first* heart sound occurs at this time. The second sound is due to closure of the aortic and pulmonary valves. and occurs at about the same time as the T wave.

30 (a) **False** It shows atrial fibrillation. (Sinus arrhythmia is a changing heart rate during the respiratory cycle due to autonomic influences on the sino-atrial node.)
   (b) **False** It shows a mixture of ventricular extrasystoles (beginning and end) and normal complexes (middle).
   (c) **True**
   (d) **True** P waves are seen to be separate from QRS complexes.
   (e) **True** This person is suffering from complete heart block. and the ventricular pacemakers are only slightly affected by catecholamines.

31 (a) **True**
   $$\text{Cardiac output (l/min)} = \frac{\text{Oxygen consumption (ml/min)}}{\text{A-V difference (ml } O_2\text{/l blood)}}$$
   (b) **True**
   (c) **False**
   (d) **True** This needs to be genuinely mixed. i.e. from right ventricle or pulmonary artery. Peripheral venous blood will not do.
   (e) **False**

32 During prolonged exercise the following cardiovascular adjustments take place.
  (a) Muscle blood flow increases partly due to increased parasympathetic vasodilator nerve discharge.
  (b) Muscle blood flow increases partly as a consequence of the local release of vasodilator metabolites.
  (c) Skin blood flow increases.
  (d) Pulmonary vascular resistance decreases.
  (e) The increase in cardiac output observed on exercise is largely brought about by the Starling mechanism.

33 If you attempt to expire forcibly against a closed glottis, the intrathoracic pressure may rise as high as +100 mm Hg (13·3 kPa). Which of the following will be sustained effects of this manoeuvre?
  (a) A rise in intratracheal pressure.
  (b) A rise in right ventricular output.
  (c) A reduction in left ventricular output.
  (d) A fall in systemic arterial pressure.
  (e) A fall in heart rate.

*34 The following are statements about the clinical assessment of the cardiovascular system:
  (a) Atrial fibrillation results in an irregular radial pulse.
  (b) A pulse rate of 40/min would suggest complete heart block.
  (c) Sphygmomanometry by the palpatory method is useful for estimating the patient's diastolic blood pressure.
  (d) An essential preliminary to sphygmomanometry by the auscultatory method is to place the bell of the stethoscope so that the arterial pulse can be clearly heard before the cuff is inflated.
  (e) The jugular vein normally fills to about 20 mm above the sternal notch when the patient is in the sitting position.

*35 Cyanosis is a dusky bluish colouration of mucous membranes and/or skin which occurs when the blood there contains 50 g/l or more of deoxygenated haemoglobin. This *could* occur as a result of:
  (a) Reduction in the pulmonary diffusing capacity for oxygen.
  (b) Shunting of blood from the left side of the heart to the right (e.g. through a patent foramen ovale).
  (c) Too high a ventilation/perfusion ratio in the lungs.
  (d) Peripheral vasoconstriction.
  (e) Increased red cell count (polycythaemia).

*36 Acute left ventricular failure is accompanied by the following:
  (a) Increase in pulmonary blood volume.
  (b) Normal jugular venous pressure.
  (c) Generalised vasodilatation.
  (d) Ankle oedema.
  (e) Raised systemic arterial pressure.

32 (a) **False** Definitely not: there is no general peripheral distribution of parasympathetic vasodilator nerves.
  (b) **True**
  (c) **True** This helps to reduce body temperature which is raised by the heat
produced by active muscle.
  (d) **True** As cardiac output increases (see answer to 28(c)).
  (e) **False** End-diastolic volume may increase during exercise in the erect posture.
but the increase in cardiac output is due mainly to the effect of the
sympathetic nervous system on the heart.

33 (a) **True**
  (b) **False** Right ventricular output falls because high intrathoracic pressure
greatly reduces venous return.
  (c) **True** This results from the fall in right ventricular output.
  (d) **True** This results from (c).
  (e) **False** Fall of systemic blood pressure leads to compensatory increase in
heart rate.

34 (a) **True**
  (b) **True** Note however that very fit young people may have resting heart rates
as low as this.
  (c) **False** Only the systolic pressure can be estimated by this method.
  (d) **False** No sound can be heard until the cuff is inflated sufficiently to cause
partial occlusion of the vessel.
  (e) **False** That would indicate a raised venous pressure.

35 (a) **True**
  (b) **False** The shunt is *left* to *right*. (This eventually leads to heart failure but
initially the blood is fully oxygenated.)
  (c) **False**
  (d) **True** Slow flow leads to removal of more oxygen from blood.
  (e) **True** Slow flow again, but this time due to increased blood viscosity.

36 (a) **True** The rise in pulmonary vascular pressure may lead to pulmonary
oedema.
  (b) **True** Jugular venous pressure is raised in *right* ventricular failure.
  (c) **False**
  (d) **False** This follows chronic right ventricular failure.
  (e) **False** Left ventricular output is decreased.

*37 The following might be expected to occur in a patient suffering from right ventricular failure of the heart:
(a) Raised central venous pressure.
(b) Enlargement of the liver.
(c) Pulmonary oedema.
(d) Reduced systemic arterial blood pressure.
(e) Decreased aldosterone plasma concentration.

*38 A blood pressure of 180/120 mm Hg (24/16 kPa) was found in a patient aged 35. The following are statements about this:
(a) Since the blood pressure is raised there must be a corresponding rise in cardiac output.
(b) Renal artery stenosis could produce this finding.
(c) Phaeochromocytoma could produce this finding.
(d) The figures quoted happen to be at the upper end of the range of blood pressures found in the healthy population of that age group.
(e) The left ventricular stroke work is increased because of the high blood pressure.

*39 Peripheral circulatory failure due to haemorrhage is accompanied by:
(a) Raised haematocrit.
(b) Rapid heart rate.
(c) Cold pale clammy skin.
(d) Reduction in central venous pressure.
(e) Increased reticulocyte count.

40 Which of the following are consequences of breathing 5% CO_2 in air?
(a) There is a rise in the PCO_2 of mixed venous blood.
(b) The pH of arterial blood increases.
(c) There is an increase in alveolar ventilation.
(d) The oxygen dissociation curve of haemoglobin is shifted to the left.
(e) There is a decrease in cerebral blood flow.

37 (a) **True**
   (b) **True**
   (c) **False** That requires raised pressure in the pulmonary capillaries, which is a feature of *left* heart failure.
   (d) **False**
   (e) **False** The aldosterone level usually rises in chronic failure.

38 (a) **False** Raised blood pressure could be due to peripheral vasoconstriction from a variety of causes.
   (b) **True** Renin is released from the affected kidney. This leads to hypertension through the renin–angiotensin–aldosterone mechanism.
   (c) **True** This is a tumour of the adrenal medulla.
   (d) **False**
   (e) **True**

39 (a) **False** The haematocrit falls as extracellular fluid moves from the interstitium into the vascular compartment.
   (b) **True** Due to increased sympathetic activity.
   (c) **True** This is due to a combination of vasoconstriction and sweating, both of which are also consequences of sympathetic activity.
   (d) **True** Central venous pressure is useful for monitoring extracellular fluid volume.
   (e) **False** This will occur in a few days if the patient survives.

40 (a) **True**
   (b) **False** Increased PCO_2 results in an increased carbonic acid concentration in the blood and this will cause a fall of pH.
   (c) **True**
   (d) **False** The curve is shifted to the right–i.e. Hb carries less O_2 at a given PCO_2.
   (e) **False** Cerebral blood flow is increased. Arterial blood PCO_2 is the main factor controlling cerebral blood flow.

41 The following are statements about the consequences of voluntarily increasing ventilation three-fold. (Note that the isoelectric point for plasma proteins is about pH 5.)
(a) The alveolar PO_2 trebles.
(b) The plasma proteins become more ionised.
(c) The plasma ionised calcium level increases.
(d) A fall in the plasma bicarbonate concentration in arterial blood.
(e) A brief period of apnoea immediately afterwards.

42 The diagram shows the relation between pulmonary ventilation and alveolar PCO_2 at two different values of alveolar PO_2. Are the following statements correct deductions from the data in the graph?

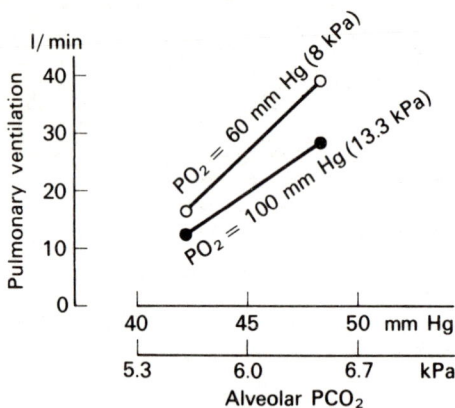

(a) There is a direct relationship between pulmonary ventilation and alveolar PCO_2.
(b) There is an inverse relationship between pulmonary ventilation and alveolar PO_2.
(c) The increment in pulmonary ventilation produced by a given rise in alveolar PCO_2 is less when the alveolar PO_2 is raised from 60 to 100 mm Hg (from 8 to 13 kPa).
(d) The straight line joining the data points for PO_2 = 60 mm Hg (open circles) is fitted by the equation:
$$\text{Pulmonary ventilation} = 4 \times \text{alveolar } PCO_2$$
$$\text{(l/min)} \qquad\qquad \text{(mm Hg)}$$
(e) When pulmonary ventilation increases, the alveolar PCO_2 rises.

43 The following measurements were made on a patient:
    Tidal volume 500 ml, Dead space 100 ml,
    Rate of breathing 15/min, Cardiac output 7 litres/min.
Are these statements correct?
(a) The pulmonary ventilation is 7·5 litres.
(b) The subject is likely to be under 8 yrs of age.
(c) The alveolar ventilation is 6 litres/min.
(d) The ventilation/perfusion ratio is the pulmonary ventilation divided by the cardiac output.
(e) The ventilation/perfusion ratio is 0·86.

41 (a) **False** Maximum PO_2 in the airways if air is being breathed is 21% of atmospheric pressure minus SVP_{H_2O} at body temperature – e.g. $0 \cdot 21 \times (760-47) = 150$ mm Hg. The value will be lower still in the alveoli because of the PCO_2.

   (b) **True** Hyperventilation will raise the plasma pH (respiratory alkalosis), thus taking the plasma proteins further from their isoelectric point.

   (c) **False** Although the total plasma $[Ca^{2+}]$ is unaffected, the *proportion* in ionized form decreases because of increased binding of calcium to various substances when pH rises. This may lead to tetany.

   (d) **True** Bicarbonate concentration falls because the CO_2 concentration falls, but note that the *ratio* $\dfrac{[HCO_3^-]}{[CO_2]}$ rises.

   (e) **True** Apnoea means absence of breathing; this is a consequence of the lowered arterial PCO_2.

42 (a) **True** At a given PO_2, increase of PCO_2 is associated with an increase in pulmonary ventilation.

   (b) **True** At a given PCO_2, increase of PO_2 is associated with a decrease in pulmonary ventilation.

   (c) **True** Slope of line is reduced.

   (d) **False** The slope is given correctly, but the equation should be $PV = (4 \times PCO_2) - C$ (value of C is about $1 \cdot 5$ l/min).

   (e) **False** The graph shows pulmonary ventilation as the dependent variable and PCO_2 as the independent variable. These cannot be transposed when dealing with homeostatic mechanisms.

43 (a) **False** It should be $7 \cdot 5$ litres *per minute*. (Errors of this kind are often missed by intelligent people: the moral of the story is that you must always check the units.)

   (b) **False** They are typical values for a 70 kg man.

   (c) **True** $15 \times (500 - 100) = 6000$ ml/min.

   (d) **False** It should be *alveolar* ventilation divided by cardiac output.

   (e) **True** Follows from (d).

44 In the experiment illustrated below, the respiratory responses to hypercapnia and
hypoxia are shown before and after denervation of the carotid and aortic bodies.
Which of the following are correct statements on the basis of the experimental
results?

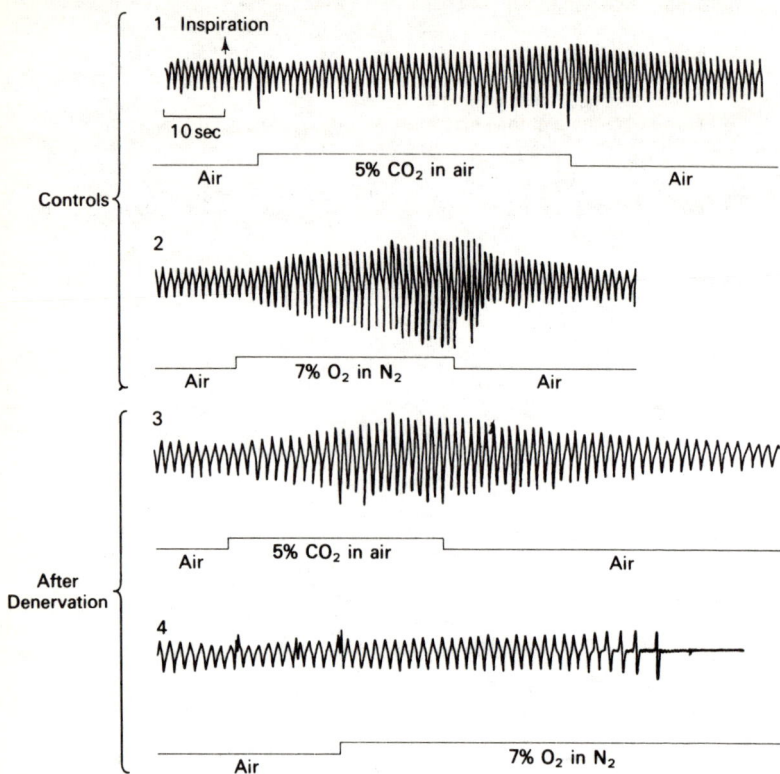

(a) Peripheral chemoreceptors (aortic and carotid bodies) are required for the
respiratory response to CO_2 excess.
(b) There are chemoreceptors other than those in the aortic and carotid bodies.
(c) Peripheral chemoreceptors are required for the response to O_2 lack.
(d) The denervation results in slowing of the respiratory rate.
(e) The effects seen in (4) are sometimes observed in normal humans (i.e. with
chemoreceptors intact).

44 (a) **False** There is still a response to breathing 5% CO_2 after denervation (Graph 3).
   (b) **True** Follows from (a).
   (c) **True** The response to breathing 7% O_2 after denervation is a depression of respiration (Graph 4).
   (d) **True** Probably due to cutting vagal afferents along with the chemoreceptor nerves, thus removing inhibitory effects on inspiration.
   (e) **False** Although some humans do not hyperventilate in response to a slight reduction in inspired PO_2, hyperventilation is always observed at values below 10%.

*45 The diagrams show spirometer tracings obtained from an adult subject in whom tidal volume, FEV_1 and vital capacity were measured before and after inhalation of isoprenaline.

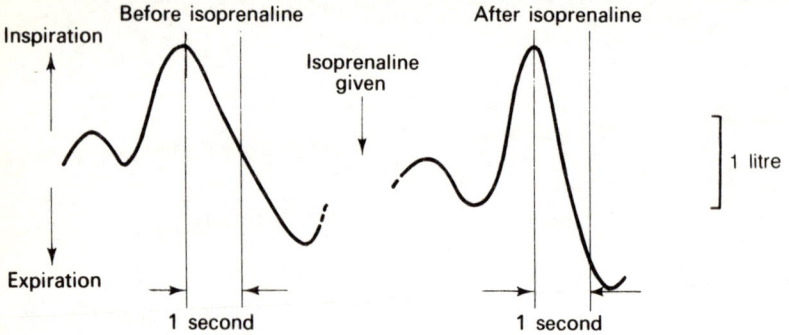

(a) Before isoprenaline, the FEV_1 was about 70% of vital capacity.
(b) The tracing obtained before isoprenaline was given was within normal limits.
(c) The vital capacity was increased by about 20% after inhalation of isoprenaline.
(d) Expiratory flow rate after inhalation of isoprenaline was consistent with a fall of airway resistance.
(e) These tracings are typical of those expected from an asthmatic patient.

46 These graphs are volume–pressure plots for the lungs during inspiration and expiration. Compared with the normal situation illustrated by graph A, the other graphs could be produced by:

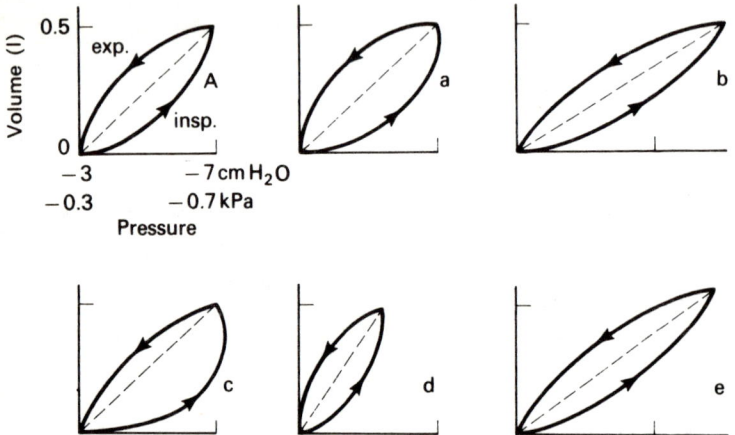

(a) An increased viscous work of breathing.
(b) A decreased lung compliance.
(c) An increase in the rate of inhalation.
(d) A diminished tidal volume.
(e) An increased elastic work of breathing.

45 (a) **False** FEV_1 was about 40% of vital capacity.
   (b) **False** FEV_1 should normally exceed 80% of vital capacity.
   (c) **True**
   (d) **True** Expiration occurred faster because isoprenaline relaxes smooth muscle of the airways.
   (e) **True**

46 (a) **True** Area of loop is increased.
   (b) **True** Bigger pressure change required for same volume change.
   (c) **True** Work of inspiration for same change of volume has increased.
   (d) **False** Tidal volume is same as in *A*.
   (e) **True** Area between dashed line and ordinate (= elastic work) is greater than in *A*.

47 The following are statements about the surface-active material (surfactant) lining the lung alveoli:
  (a) The surfactant increases the surface tension of the film of liquid lining the alveoli.
  (b) Surfactant increases lung compliance.
  (c) The surface tension of fluid containing surfactant increases as the surface area of the fluid decreases.
  (d) In the absence of normal surfactant there will be a greater tendency for alveoli to collapse.
  (e) An indirect consequence of absence of normal surfactant may be a fall in systemic arterial pH.

*48 Analyses of gases in alveolar air and systemic arterial blood were made on a patient, and were:
      Alveolar air PO_2 102 mm Hg (13·6 kPa); Systemic arterial blood PO_2 70 mm Hg (9·3 kPa).
    Are the followng statements based on these values correct?
  (a) They are typical values for a healthy patient.
  (b) They could be explained by a reduced diffusing capacity for oxygen.
  (c) The values are typical for a healthy person who lives at high altitude.
  (d) They could be explained by a shunt of deoxygenated blood into systemic arterial blood.
  (e) The patient may have been hypoventilating.

*49 The following are statements about the ventilation and perfusion of alveoli with blood:
  (a) Overperfusion of normally ventilated lungs leads to a lowered PCO_2 in systemic arterial blood.
  (b) Overperfusion of normally ventilated lungs leads to a lowered PO_2 in systemic arterial blood.
  (c) Overventilation of some alveoli that are normally perfused can compensate for changes in systemic arterial PCO_2 caused by overperfusion of other alveoli.
  (d) If alveoli are underperfused and normally ventilated the blood leaving them will have a normal oxygen content.
  (e) If some alveoli are both underperfused and underventilated, the gas tensions in the pulmonary capillary blood could well have normal values.

50 The following statements concern pulmonary blood flow:
  (a) The pulse pressure in the pulmonary artery is about the same as that in the aorta.
  (b) If the blood pressure in pulmonary capillaries is 5 mm Hg (0·7 kPa) above the intra-alveolar pressure, fluid will pass into the alveolar air spaces.
  (c) The ventilation/perfusion ratio is the same in all parts of the lung in standing man.
  (d) During prolonged bouts of coughing the venous return to the right side of the heart is reduced.
  (e) Deep inspiration will increase the capacity of the pulmonary capillaries.

47 (a) **False** Surfactant decreases surface tension.
   (b) **True** It reduces the force required to overcome surface tension effects when alveoli expand.
   (c) **False** Because the surfactant remains at the water air interface, the space between surfactant molecules decreases as the surface area is reduced; this is equivalent to raising its concentration, which lowers surface tension.
   (d) **True** This is a consequence of the raised surface tension.
   (e) **True** The greater tendency of the alveoli to collapse can lead to inadequate ventilation and a rise in alveolar PCO_2 – and hence a fall in pH of the systemic arterial blood.

48 (a) **False** The alveolar value is normal but the systemic arterial PO_2 should be much closer to it.
   (b) **True**
   (c) **False** The alveolar PO_2 would be lower, and in any case this would not account for the large difference between alveolar and systemic arterial blood values.
   (d) **True**
   (e) **False** See comment for (c).

49 (a) **False** The ventilation/perfusion ratio will be lowered. The blood in the lungs will not equilibrate with alveolar air, so that systemic arterial PCO_2 (P_aCO_2) will rise. It cannot fall in these circumstances.
   (b) **True** A similar explanation to (a), leading to a fall in P_aO_2.
   (c) **True** Regional overventilation can reduce CO_2 in blood leaving the overventilated alveoli. This reduction can compensate for the increase of CO_2 in blood from overperfused alveoli. Note that overventilation will not markedly alter the oxygen content since Hb is saturated at the normal alveolar PO_2 of 100 mm Hg.
   (d) **True** The blood will have equilibrated with the alveolar air.
   (e) **True** Abnormal distribution of the inspired gas in the alveoli (due for example to compliance or resistance changes) does not necessarily lead to abnormal gas exchange. Compensatory mechanisms can match blood flow to the ventilation.

50 (a) **False** Systolic and diastolic pressures in the pulmonary artery are about 1/6 of those in the aorta and so is the pulse pressure.
   (b) **False** The colloid osmotic pressure of plasma is enough to prevent this.
   (c) **False** Blood flow is relatively less in the upper parts of the lung than the lower regions, but this is not matched by the differences in degree of ventilation.
   (d) **True** Greatly increased intrapleural pressure can collapse great veins within the thorax.
   (e) **True** This results from the increased transmural pressure caused by the subatmospheric extravascular (intrapleural) pressure.

51 The following are statements about thirst:
   (a) Drinking may be provoked by electrical stimulation of certain regions of the hypothalamus.
   (b) Loss of blood volume (e.g. haemorrhage) is not accompanied by thirst if there is no change in plasma osmolarity.
   (c) Thirst is produced by a rise in plasma osmolarity even if the blood volume is normal.
   (d) If a dehydrated person is allowed access to water, thirst and drinking continue unabated until the plasma osmolarity is restored to normal.
   (e) The sensation of thirst is related to the rate of resting salivation.

52 The following are statements about glomerular filtration:
   (a) The glomerular filtrate is produced by essentially the same mechanisms as interstitial fluid.
   (b) Glomerular filtrate has the same composition às lymph collected from the thoracic duct.
   (c) Blood in the efferent glomerular arteriole (i.e. the one carrying blood away from the glomerulus) is more viscous than blood in the afferent arteriole.
   (d) The glomerular filtration rate (GFR) is directly proportional to the systemic arterial blood pressure.
   (e) The glomerular filtration rate is the main factor determining the rate of urine production.

53 If the concentration of a substance X in the plasma is P_X (mg/ml) and in the urine is U_X (mg/ml), and the volume of urine produced per minute is V (ml), which of the following statements are correct?
   (a) The rate of excretion of substance X is $U_X.V$ (mg/min).
   (b) The quantity $U_X.V/P_X$ is the minimum volume of plasma from which the kidneys could have obtained the amount of X excreted per minute.
   (c) If X is filtered at the glomerulus and neither secreted nor absorbed in the renal tubules, then $U_X.V/P_X$ is the volume of plasma that passes through the kidney in one minute.
   (d) If X is inulin, then $U_X.V/P_X$ is the volume of glomerular filtrate produced per minute.
   (e) If the ratio U_X/P_X exceeds the ratio U_{inulin}/P_{inulin} then substance X must be secreted by the renal tubules.

51 (a) **True**
   (b) **False** Thirst is a prominent feature of a sudden reduction in blood volume.
   (c) **True**
   (d) **False** Thirst is initially slaked by drinking an amount of water that is insufficient to restore plasma osmolarity to normal; subsequently thirst returns, and drinking occurs once more.
   (e) **True**

52 (a) **True** Both are formed by ultrafiltration.
   (b) **False** It contains only traces of protein, whereas lymph contains appreciable amounts.
   (c) **True** The loss of 1/6 of the plasma volume as glomerular filtrate will result in increased concentrations of red cells (i.e. raised haematocrit) and plasma proteins.
   (d) **False** Glomerular capillary pressure and glomerular filtration rate are largely independent of changes in systemic arterial pressure in healthy kidneys because of autoregulation of blood flow.
   (e) **False** The rate of urine production in humans is dominated by tubular function, not by GFR.

53 (a) **True**
   (b) **True** This is a definition of the concept of renal 'clearance' of X.
   (c) **False** It is the volume of glomerular filtrate in one minute. If X were a substance such as PAH or diodrast, secreted as well as filtered so that the kidney removed all X going to it, then the clearance gives renal plasma flow.
   (d) **True**
   (e) **True** Because inulin is only filtered.

54 This diagram shows how the rates of filtration (F), secretion (S), and excretion (E) of para-amino-hippuric acid (PAH) vary with its plasma concentration (P_{PAH}).

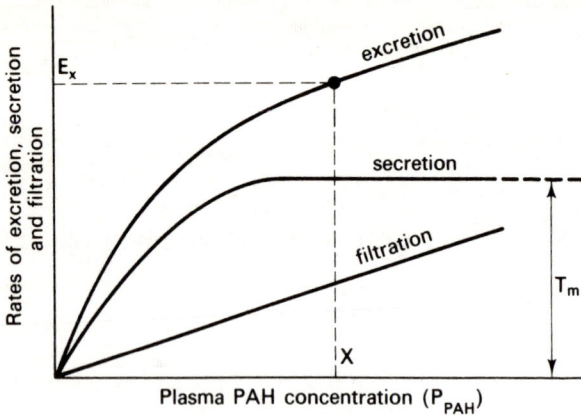

Plasma PAH concentration (P_{PAH})

(a) The rate of filtration of PAH is directly proportional to its plasma concentration.
(b) At all values of P_{PAH} the rate of excretion of PAH is given by
$$E = F + S$$
(c) T_m is the maximum tubular reabsorptive capacity for PAH.
(d) The rate of filtration of PAH is given by
$$F = k.P_{PAH}$$
where k is the slope of the filtration curve in the above diagram.
(e) The rate of excretion of PAH at plasma concentration, X, is given by
$$E_x = T_m - k.X$$

55 In healthy adults, plasma inorganic phosphate concentration lies between $0 \cdot 6$ and $1 \cdot 5$ mmol/l and inulin clearance is typically 120 ml/min. Phosphate appears in the urine when plasma levels exceed about $0 \cdot 9$ mmol/l.
(a) T_m phosphate is about $0 \cdot 11$ mmol/min.
(b) If the plasma phosphate concentration is 1 mmol/l, the amount of phosphate excreted in the urine will be about $0 \cdot 12$ mmol/min.
(c) Parathormone tends to reduce T_m phosphate.
*(d) A low value for the ratio T_m phosphate: inulin clearance can lead to hypophosphataemia.
*(e) A low T_m phosphate combined with a high inulin clearance is commonly found in renal failure.

54 (a) **True** Graph is a straight line through the origin.
   (b) **True**
   (c) **False** T_m (as illustrated here) is the maximum tubular *secretory* capacity. (A nasty catch question, but it emphasizes the need to read the statement word by word before making a decision.)
   (d) **True** This is an algebraic statement of (a).
   (e) **False** $E_x = T_m + k.X$.

55 (a) **True** Inulin clearance is a measure of the glomerular filtration rate so the delivery of phosphate to the renal tubules (filtered load) is:
   GFR x [phosphate] = $0 \cdot 12$ × $0 \cdot 9$ = $0 \cdot 108$
                 l/min   mmol/l   mmol/min
   Phosphate appears in the urine when the plasma concentration exceeds $0 \cdot 9$ mmol/l, because the filtered load then exceeds the T_m for phosphate reabsorption.
   (b) **False** Amount excreted = filtered load   − amount reabsorbed (T_m)
                        $(1 × 0 \cdot 12)$ − $0 \cdot 108$
                        mmol/l l/min   mmol/min
                   = $0 \cdot 012$ mmol/min
   (c) **True** This is one of the actions of parathormone.
   (d) **True** Significant hypophosphataemia is generally seen only if T_m falls. This may be in response to a raised parathormone level ('secondary hypophosphataemia') or a direct failure of tubular reabsorption of phosphate ('primary hypophosphataemia').
   (e) **False** In chronic renal failure GFR falls and T_m falls roughly in parallel. This is widely believed to indicate that the number of active nephrons is reduced.

56 The following are statements about renal tubules:
   (a) At the tip of the loop of Henle in the renal medulla, the osmolarity of the tubular contents is several times that of the glomerular filtrate.
   (b) The walls of the ascending limb of the loop of Henle are freely permeable to water.
   (c) The fluid entering the distal convoluted tubule is hypotonic with respect to plasma.
   (d) The main force responsible for water reabsorption from the collecting tubule is the sub-atmospheric pressure in the surrounding interstitial space.
   (e) The permeability of the collecting tubule to water is under the control of aldosterone.

*57 A loss of water in excess of NaCl can occur in certain situations.
   (a) When this happens, glomerular filtration rate is usually increased.
   (b) Diabetes mellitus would give rise to this condition.
   (c) Diabetes insipidus would give rise to this condition.
   (d) Addison's disease would give rise to this condition.
   (e) Severe sweating would give rise to this condition.

*58 In a primary dehydration (also called a hypertonic contraction), there is a loss of water in excess of salt; some features of this condition are:
   (a) The osmolarity of the intracellular fluid is increased.
   (b) The intracellular fluid volume remains unchanged.
   (c) There is a low rate of production of concentrated urine.
   (d) The rate of salt loss in the urine is increased.
   (e) The body temperature often rises.

*59 In a secondary dehydration (also called a hypotonic contraction), there is a loss of salt in excess of water; some features of this condition are:
   (a) A decrease in intracellular fluid volume.
   (b) The patient complains of severe thirst.
   (c) Haemoconcentration occurs (raised haematocrit and plasma protein concentration).
   (d) There is a good flow of dilute urine.
   (e) Marked peripheral vasoconstriction may occur.

*60 Ankle oedema accompanies the following diseases. (Note that it is assumed that each is uncomplicated by other diseases.)
   (a) Renal failure (nephrotic syndrome).
   (b) Kwashiorkor.
   (c) Heart failure.
   (d) Myxoedema.
   (e) Hypertension.

56 (a) **True** This results from counter current multiplier.
   (b) **False** They are relatively impermeable to water.
   (c) **True** Due to (b) and active removal of NaCl from lumen of ascending limb of loop of Henle.
   (d) **False** The force is osmotic, due to high osmolarity of the interstitial fluid, compared with that of the tubular contents.
   (e) **False** The permeability is controlled by antidiuretic hormone (ADH).

57 (a) **False** Glomerular filtration is not regulated to compensate for disorders of salt and water balance.
   (b) **True** Because diabetes mellitus can result in an osmotic diuresis.
   (c) **True** Lack of ADH results in excretion of a hypotonic urine.
   (d) **False** Loss of salt in excess of water is the primary problem when there is a deficiency of adrenal cortical hormones (aldosterone in partcular).
   (e) **True** Sweat contains NaCl, but is less concentrated than the interstitial fluid.

58 (a) **True** Osmolarity of all fluid compartments is increased.
   (b) **False** Volume of all fluid compartments is decreased.
   (c) **True** Due to high rate of secretion of ADH by posterior pituitary.
   (d) **True** Due to reduced rate of secretion of aldosterone by adrenal cortex.
   (e) **True** This culminates in 'dehydration fever' due to failure of circulation to dissipate heat when blood volume is severely reduced.

59 (a) **False** The intracellular fluid volume is increased because of an osmotic movement of water from ECF when its osmolarity decreases.
   (b) **False**
   (c) **True** Severe because any loss of fluid to the exterior is aggravated by loss to ICF.
   (d) **True** Consequence of (a).
   (e) **True** Required for maintenance of blood pressure in face of greatly diminished blood volume.

60 (a) **True** Colloid osmotic pressure falls due to renal loss of plasma proteins.
   (b) **True** Dietary deficiency of essential amino acids results in lack of plasma proteins.
   (c) **True** This results mainly from expansion of the ECF due to excessive aldosterone production, but it is aggravated by raised venous pressures in severe forms of congestive cardiac failure.
   (d) **False** A thickening of the subcutaneous tissue is seen in thyroid deficiency as a result of changes in the ground substance.
   (e) **False** Oedema is not a feature of hypertension unless there are complications (e.g. heart failure).

61 The following are statements about the acidification of urine:
   (a) H^+ is secreted into the urine by the cells lining the distal tubules.
   (b) K^+ is normally reabsorbed from the tubular fluid in exchange for H^+.
   (c) The H^+ reacts with the NaH_2PO_4 in the tubular fluid to give Na_2HPO_4.
   (d) When there is a large load of H^+ to be excreted, most of it appears in the urine in the form of ammonium salts.
   (e) H_2CO_3 is present in the urine in very high concentration compared with plasma.

*62 The diagram shows the relation between plasma bicarbonate concentration and the pH of blood at three different values of PCO_2 (60, 40 and 20 mm Hg) (8, 5·3 and 2·7 kPa). N represents the blood picture in a normal person.

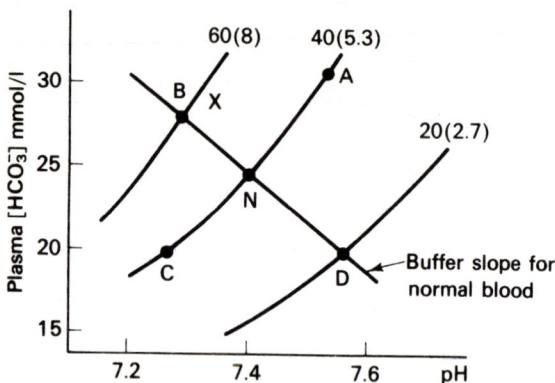

   (a) A movement along the line NB occurs in an uncompensated respiratory acidosis.
   (b) The equation of line ANC is $pH = pK' + \log_{10} \dfrac{[HCO_3^-]}{a.PCO_2}$
       where a = solubility coefficient for CO_2 and PCO_2 = 40 mm Hg (5·3 kPa).
   (c) A movement along the line NA could be produced by hyperventilation.
   (d) Point X could be reached in a partially compensated metabolic acidosis.
   (e) The slope of the line BND will be reduced if the haemoglobin concentration in the blood is decreased.

61 (a) **True** It is also secreted in the proximal tubules.
   (b) **False** Na^+ is reabsorbed in exchange for H^+ or K^+.
   (c) **False** Other way around.
   (d) **True** The urine smells of ammonia under these conditions.
   (e) **False** The formation of H_2CO_3 occurs by combination of secreted H^+ with HCO_3^- in the tubular fluid, but it is followed by breakdown to $CO_2 + H_2O$, and the CO_2 diffuses back into the cells.

62 (a) **True** In the absence of compensation, the pH falls and the $[HCO_3^-]$ rises.
   (b) **True** This is the Henderson–Hasselbalch equation. The curvature of the line ANC reflects the logarithmic relation between pH and $[HCO_3^-]$.
   (c) **False** In hyperventilation the PCO_2 falls. In the absence of any renal compensation, a respiratory alkalosis produces a movement along line ND (i.e. the opposite of a respiratory acidosis).
   (d) **False** A metabolic (i.e. non-respiratory) acidosis produces a movement along line NC, and the locus reached as a result of compensation will lie below and to the right of C because of the fall in PCO_2 (due to hyperventilation) and a fall in $[HCO_3^-]$ (due to hyperventilation and increased excretion of HCO_3^- in the urine).
   (e) **True** Haemoglobin is the main buffer in the blood and its concentration is the main factor determining the slope of this line.

*63 At the beginning of an illness a sample of a patient's systemic arterial blood had
    the following composition: pH 7·2, PCO_2 38 mm Hg (5·1 kPa), plasma [HCO_3^-]
    14·3 mmol/l. The illness continued and two days later the situation was: pH 7·36,
    PCO_2 33 mm Hg (4·4 kPa), plasma [HCO_3^-] 18 mmol/l.
    (a) The second blood sample was within normal limits.
    (b) A possible diagnosis was respiratory acidosis.
    (c) A possible diagnosis was non-respiratory alkalosis.
    (d) Respiratory compensation must have taken place between the taking of the
        first and second blood samples.
    (e) Renal compensation must have taken place between the time of the first and
        second blood samples.

*64 Persistent vomiting leads to a complicated disorder of acid-base balance, because
    there is a loss of both H^+ and K^+ from the body.
    (a) The loss of H^+ results in a non-respiratory alkalosis.
    (b) The loss of K^+ results in an intracellular alkalosis.
    (c) The plasma HCO_3^- concentration will be lower than normal.
    (d) Respiration will be depressed.
    (e) An acid urine will be produced.

 65 A pouch of the stomach may be made such that it is denervated but still has its
    normal blood supply. Gastric juice may be collected from such a pouch in the
    conscious dog. The following are statements about secretion from the pouch:
    (a) There is a resting secretion.
    (b) A normal secretory response to the sight of food is produced.
    (c) Secretion starts about $\frac{1}{2}$ hour after a meal is taken by mouth.
    (d) Distension of the pouch increases secretion.
    (e) If the pouch is moved to a different part of the body and a new blood supply
        established, secretion occurs in response to a meal.

*66 Total gastrectomy would be expected to lead to the following:
    (a) Haemodilution after meals.
    (b) Vitamin B_{12} malabsorption.
    (c) Grossly reduced iron absorption.
    (d) Malabsorption of protein.
    (e) Impaired fat absorption.

63 (a) **False** The pH is within normal limits. but the PCO_2 and HCO_3^- are not.
   (b) **False** PCO_2 is not raised above normal value (~ 40 mm Hg) ($5 \cdot 3$ kPa).
   (c) **False** The pH of the first blood sample is below normal.
   (d) **True** Respiratory compensation would explain the reduced PCO_2 but renal compensation is required to account for the rise of HCO_3^- concentration.
   (e) **True** As shown by the rise in HCO_3^- concentration.

64 (a) **True** Non-respiratory alkalosis is often called metabolic alkalosis.
   (b) **False** The loss of K^+ from the body depletes the intracellular stores. and H^+ replaces the lost K^+; this results in an intracellular *acidosis*.
   (c) **False** It is raised in a non-respiratory alkalosis (but lowered in a respiratory alkalosis – think about that).
   (d) **True** This occurs because of the raised plasma pH.
   (e) **True** This is not the renal response you would expect to an alkalosis; it results from (b). and it is called paradoxical aciduria.

65 (a) **True** There is normally a low level of circulating gastrin.
   (b) **False** This response is neurally mediated via the vagus.
   (c) **True** The delayed response is humorally mediated. and is due to gastrin release.
   (d) **True** The secretory response to distension is largely a local response.
   (e) **True** Humoral influences are still effective because they act via the blood stream.

66 (a) **False** Haemoconcentration is more likely. After gastrectomy the presence of large volumes of hypertonic digestion products in the duodenum draws water into the gut lumen (dumping syndrome).
   (b) **True** There would be a lack of intrinsic factor.
   (c) **False** It might be reduced slightly due to lack of conversion of Fe^{3+} to Fe^{2+}.
   (d) **True** There is submaximal stimulation of duodenal hormone release (e.g. secretin and CCK) and hence submaximal stimulation of pancreatic exocrine stimulation.
   (e) **True** See answer to (d).

67 The following are statements about pancreatic function:
   (a) Stimulation of the vagus produces a rapid secretion of watery juice from the pancreas.
   (b) In cross-circulation experiments when acid is introduced into the duodenum of one animal, pancreatic secretion occurs in the other.
   (c) Pancreatic secretion contains enzymes which break down polysaccharides.
   (d) Pancreatic acini contain trypsin.
   (e) The hormone responsible for provoking the enzyme-rich pancreatic secretion also causes contraction of the gall bladder.

68 The following are statements about patients lacking exocrine pancreatic secretion:
   (a) Fat digestion is normal provided bile is still produced.
   (b) Protein digestion is inefficient and loss of weight tends to occur.
   (c) A bleeding tendency may be present.
   (d) Excessive water loss takes place via the gut primarily because of inadequate Na^+ absorption in the ileum.
   (e) Fasting blood sugar is elevated.

69 The following are statements about bile:
   (a) Bile salts are derived from waste products of haemoglobin breakdown.
   (b) Reabsorption of bile salts from the intestine leads to further secretion of bile.
   (c) A certain concentration of bile salt is required before normal fat absorption can take place.
   (d) Bile is concentrated in the gall bladder.
   (e) Active transport of NaCl out of the gall bladder is the mechanism by which the bile is concentrated.

*70 The following are statements about jaundice.
   (a) In obstructive jaundice the faeces are pale and the urine is dark.
   (b) Obstructive jaundice is commonly the result of gall stones blocking the cystic duct.
   (c) Hepatocellular jaundice leads to an excessive excretion of bilirubin by the kidney.
   (d) Haemolytic jaundice does not lead to changes in the faecal content of stercobilinogen.
   (e) The production of bilirubin from haemoglobin does not take place in liver failure.

71 The following are statements about fat absorption:
   (a) If fats are shaken with water at 37°C, the droplet size is greater when bile salt is also present.
   (b) Chylomicra are small droplets of fat found in the small intestinal lumen.
   (c) Most bile sat is absorted in the duodenum.
   (d) The breakdown products of dietary fats are resynthesised in the intestinal cells and pass into the lacteals.
   (e) Deficiencies of fat absorption can lead to poor absorption of vitamins of the B group.

67 (a) **False** That is typical of the effect of secretin. Vagal stimulation produces a scanty enzyme-rich juice.

 (b) **True** Acid stimulates the release of secretin into the blood stream from the duodenum.

 (c) **True** Amylases.

 (d) **False** Trypsin is stored and secreted in an *inactive* form, trypsinogen. Activation of trypsin in the gland causes digestion of the pancreas!

 (e) **True** It was once thought there were two local hormones – cholecystokinin and pancreozymin – but they are now known to be the same thing (CCK–PZ).

68 (a) **False** Pancreatic lipases are required in addition to bile salts to break down the triglycerides.

 (b) **True** Pancreatic proteases and peptidases are essential for protein digestion.

 (c) **True** Due to poor absorption of fats and fat-soluble vitamins including vitamin K, which is required for the production of prothrombin.

 (d) **False** Pancreatic secretion has no effect on sodium absorption.

 (e) **False** That would result from loss of endocrine function because of lack of insulin.

69 (a) **False** You are getting confused with bile pigments, which are the products of haemoglobin metabolism.

 (b) **True** This is the 'enterohepatic' circulation of bile.

 (c) **True** At a sufficient concentration of bile (the critical micellar concentration) water soluble aggregates of bile salts, free fatty acids, monoglycerides, cholesterol and fat-soluble vitamins are formed in the gut.

 (d) **True** About 10- to 20-fold.

 (e) **True** Water moves isosmotically with it.

70 (a) **True** Lack of bile makes the faeces pale, but conjugated bile is excreted in the urine.

 (b) **False** The cystic duct connects the gall bladder to the common bile duct; blockage of the cystic duct does not therefore cause jaundice.

 (c) **False** Damage to the liver leads to failure of conjugaton and no renal excretion of bilirubin can cccur.

 (d) **False** Stercobilinogen is increased because excessive haemolysis leads to excessive bile pigment excretion in the faeces.

 (e) **False** It is *conjugation* that is impaired in liver failure.

71 (a) **False** The bile salts act as emulsifying agents so droplet size is smaller.

 (b) **False** They are found in mesenteric lymphatics and in the blood stream after a fatty meal.

 (c) **False** In the terminal ileum by active transport.

 (d) **True**

 (e) **False** The fat soluble vitamins are A, D, E and K. If you cannot remember them, invent a mnemonic.

72 The following are statements about iron metabolism.
   (a) The main excretory route for iron is via cells shed from the intestinal mucosa.
   (b) Destroyed red cells provide the main immediate source of plasma iron.
   (c) Apoferritin production depends on the plasma iron concentration.
   (d) Apoferritin present in the intestinal mucosal cells prevents iron from gaining access to the circulation by combining with it.
   (e) Most of the iron in the plasma is in the free form.

73 The graphs show the transport rate of individual sugars across a membrane. Transport is occurring from a solution containing the sugar both alone and in the presence of a fixed concentration of other sugars into a solution containing no sugar. Which of the following statements are consistent with the data?

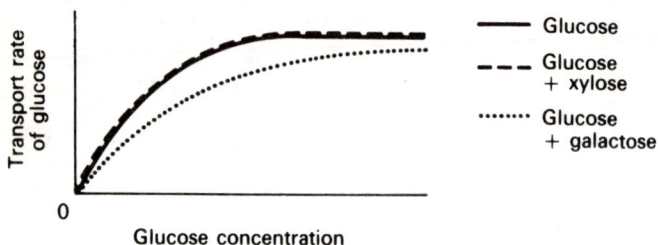

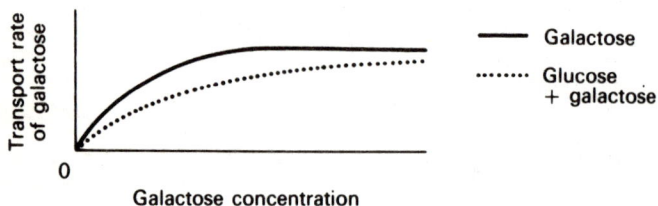

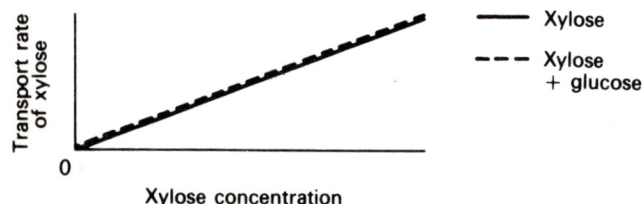

(a) Glucose is transported actively.
(b) Galactose is transported actively.
(c) Xylose is transported actively.
(d) Glucose and galactose compete for the same carrier mechanism.
(e) Glucose and xylose compete for the same carrier mechanism.

72 (a) **True**
   (b) **True**
   (c) **True**
   (d) **True** Increased plasma iron leads to the production of more apoferritin, which combines with iron and is subsequently shed in the lumen.
   (e) **False** Most of the iron is combined with a carrier protein transferrin.

73 (a) **True** A limiting rate is observed.
   (b) **True** For the same reason.
   (c) **False** Rate of transport increases in proportion to lumen concentration.
   (d) **True** Presence of one reduces the rate of transport of the other.
   (e) **False** Presence of xylose makes no difference to rate of transport of glucose.

*74 The following are statements about diseases affecting the intestine.
   (a) In gluten enteropathy the villi of the jejunum are flattened and reduced in number.
   (b) Gastrin-secreting tumours are sometimes found in the pancreas.
   (c) A failure of both pancreatic and biliary secretion is needed before steatorrhoea develops.
   (d) Cholera toxin leads to excessive secretion of NaCl and water into the intestinal lumen.
   (e) Megaloblastic anaemia is commonly associated with diseases that affect the terminal ileum (e.g. Crohn's disease).

75 The following are statements about the antidiuretic hormone (ADH):
   (a) ADH is transported from hypothalamic nuclei to the neurohypophysis via the hypophyseal portal system.
   (b) ADH increases the permeability of the distal and collecting tubules of the kidney to water.
   (c) A high rate of secretion of ADH can lead to complete cessation of urine production.
   (d) A fall in the osmolarity of the blood supplying the hypothalamus is a powerful stimulus for ADH secretion.
   (e) A fall in blood volume results in an increase in ADH secretion.

76 Aldosterone production is likely to be increased:
   (a) If the osmolarity of the extracellular fluid rises.
   (b) After a severe haemorrhage.
   (c) If plasma $[K^+]$ rises.
   (d) In hypopituitarism.
   (e) Following administration of a carbonic anhydrase inhibitor.

*77 Adrenal cortical insufficiency (Addison's disease) is often accompanied by a low plasma Na^+ concentration and raised plasma K^+ concentration, low blood glucose, pigmentation of the skin, and a low systemic arterial blood pressure:
   (a) The low plasma Na^+ concentration is due largely to failure of the distal convoluted tubules of the kidney to reabsorb sodium.
   (b) The raised plasma K^+ concentration is due to excessive haemolysis of red blood cells.
   (c) The low blood glucose is due to lack of glucocorticoid secretion.
   (d) The pigmentation of the skin is due to the decreased rate of production of ACTH.
   (e) The low blood pressure is due to the increase in extracellular fluid volume.

74 (a) **True**   Absorption is impaired.
   (b) **True**   This leads to excessive production of HCl (Zollinger–Ellison syndrome).
   (c) **False** A failure of either leads to malabsorption of fats.
   (d) **True**   Normal NaCl absorption is reversed by a serosal-to-mucosal Cl^- pump, activated by cAMP. Glucose-dependent Na^+ reabsorption is unaffected.
   (e) **True**   Vitamin B_{12} is absorbed in the terminal ileum.

75 (a) **False** It is transported via neurosecretory granules along the hypothalamo-neurohypophyseal tract. The portal system goes to the anterior pituitary (adenohypophysis).
   (b) **True**
   (c) **False** The upper limit of urine concentration is that of the interstitial concentration at the tip of the renal papillae and is about 1200 mosmol/l. The normal solute load at maximum urine concentration requires about 600 ml urine in 24 hr.
   (d) **False** A *rise* is required.
   (e) **True**   Blood volume is monitored by receptors in the walls of the great veins and the atria. A fall in blood volume causes a neurally-mediated release of ADH, in addition to the release of aldosterone. The latter then increases plasma Na^+ concentration which thus raises plasma osmolarity and augments the release of ADH.

76 (a) **False** A *decrease* in $[Na^+]$ is one of the stimuli for increased aldosterone production.
   (b) **True**   The reduction in extracellular volume is the stimulus in this situation.
   (c) **True**   The rate of production of aldosterone is directly related to the plasma $[K^+]$.
   (d) **False** Aldosterone production continues in the absence of anterior pituitary hormones, but is somewhat reduced.
   (e) **True**   The action of carbonic anhydrase in the convoluted tubules is necessary for re-uptake of bicarbonate, which is accompanied by Na^+, from the tubular fluid.

77 (a) **True**   Aldosterone affects sodium reabsorption in both proximal and distal tubules, but it is the impairment of absorption in the latter that results in the low plasma Na^+ concentration.
   (b) **False** It is a result of the failure of the aldosterone-dependent secretion of K^+ in exchange for Na^+ absorbed in the distal convoluted tubule.
   (c) **True**   Glucocorticoids promote glycogenolysis and gluconeogenesis.
   (d) **False** It is due to an increased rate of production of ACTH or of some trophic hormone closely related to it.
   (e) **False** The deficiency of aldosterone leads to a reduction of ECF volume and the fall in blood volume is partly responsible for the low blood pressure. The lack of glucocorticoids is also important because the arteriolar smooth muscle becomes less responsive to pressor influences.

*78 The following are common features of pan-hypopituitarism:
  (a) Amenorrhoea.
  (b) Increased urine output.
  (c) Excessive growth of body hair.
  (d) Decreased tolerance to cold weather.
  (e) A form of diabetes mellitus.

*79 Would you expect poor calcification of bone to occur in the following situations?
  (a) Cushing's syndrome.
  (b) Hypoparathyroidism.
  (c) Poor dietary intake of cholecalciferol (vitamin D) in black children living in northern areas of Europe.
  (d) Tumours of the thyroid C cells.
  (e) Renal disease.

80 The following are statements about the female reproductive system:
  (a) Both oestrogen and progesterone are necessary for ovulation to take place.
  (b) Oestrogen tends to inhibit the production of FSH by the anterior pituitary gland.
  (c) Fertilization of the ovum by the spermatozoon normally takes place in the uterus.
  (d) Progesterone production is largely under the control of LH.
  (e) Throughout the part of the menstrual cycle that follows ovulation, there is a slight rise in body temperature.

81 The question is about hypothalamic control of the release of hormones from the pituitary gland.
  (a) Oxytocin is formed in the posterior lobe of the pituitary gland.
  (b) Tactile stimulation of the nipple during suckling triggers the output of a hypothalamic releasing factor for oxytocin.
  (c) The hypothalamic releasing hormone called gonadotrophin releasing hormone (GnRH) reaches the anterior pituitary via portal blood vessels.
  (d) An increase in concentration of thyroid hormones in the plasma inhibits the production in the hypothalamus of thyrotrophin releasing hormone (TRH).
  (e) All the hypothalamic releasing hormones or factors that control hormone release from the pituitary are excitatory (i.e. they increase release).

78 (a) **True** Pan-hypopituitarism means loss of function of both anterior and posterior pituitary. Gonadotrophins are required for menstruation.
   (b) **True** Due to the absence of ADH.
   (c) **False** Hirsuitism is characteristic of excess corticoid secretion e.g. Cushing's syndrome.
   (d) **True** The thyroid response would be abolished because of TSH lack and the general response to stress is diminished because of ACTH lack.
   (e) **False** *Excessive* secretion of growth hormone can produce this.

79 (a) **True** High steroid output leads to poor collagen synthesis. Since collagen is being broken down continuously, albeit slowly, this leads to dissolution of the bone matrix.
   (b) **False** Poor calcification occurs in hyperparathyroidism.
   (c) **True** Vitamin D is required for normal calcium absorption. When dietary intake is inadequate, synthesis in the skin of calciferol by UV irradiation from the sun becomes important. Pigmentation reduces UV penetration and in northern areas there is little sun anyway.
   (d) **False** These produce calcitonin. Calcitonin enhances bone calcification.
   (e) **True** The active form of vitamin D is produced in the kidney. Renal disease can lead to 'renal rickets' in children and osteomalacia in adults.

80 (a) **False** FSH and LH are needed.
   (b) **True**
   (c) **False** It normally takes place in the uterine tubes.
   (d) **True** LH acts on the corpus luteum.
   (e) **True** Occasionally there is a transient fall at the time of ovulation before the sustained rise.

81 (a) **False** It is released from the posterior pituitary but is formed in the paraventricular nucleus of the hypothalamus. It travels down the axons of the paraventricular cells to the posterior lobe (neurohypophysis).
   (b) **False** It is a neural reflex, causing the release of oxytocin from the posterior pituitary.
   (c) **True** There is no neural link between the hypothalamus and the anterior pituitary (adenohypophysis). Communication is solely via the portal blood supply.
   (d) **True** The negative feedback control of thyrotrophin (TSH) and of thyroid hormone production acts at both hypothalamic and pituitary levels.
   (e) **False** Two inhibitory substances are known: (i) somatostatin (SS) is a hormone that inhibits release of somatotrophin, and (ii) an unidentified hypothalamic factor inhibits prolactin release.

*82 Which of the following would suggest a diagnosis of thyrotoxicosis?
   (a) Pulse rate 65/min.
   (b) Loss of weight.
   (c) Raised plasma triiodothyronine concentration.
   (d) Raised plasma TSH concentration.
   (e) Poor appetite.

*83 The following would be consistent with a diagnosis of diabetes mellitus:
   (a) A glucose tolerance test as in the diagram.

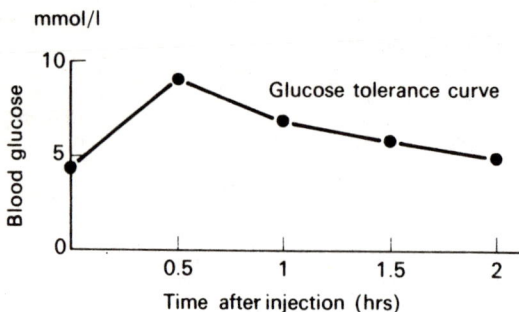

   (b) The following acid-base picture in the blood: pH 7·15; PCO_2, 35 mm Hg
       (4·6 kPa); HCO_3^-, 15 mmol/l.
   (c) Ketoaciduria.
   (d) Loss of weight.
   (e) Nocturia.

84 Net movement of a substance by simple diffusion:
   (a) *Can* occur against a concentration gradient.
   (b) Can occur against an electrochemical gradient.
   (c) Can be slowed by a factor of at least two by lowering the temperature 10°C.
   (d) Is unaffected by metabolic poisons.
   (e) Shows saturation properties.

85 Net movement of a substance by active transport:
   (a) Can occur against a concentration gradient.
   (b) Can occur against an electrochemical gradient.
   (c) Can be slowed by a factor of at least two by lowering the temperature 10°C.
   (d) Is unaffected by metabolic poisons.
   (e) Shows saturation properties.

82 (a) **False** Basal metabolic rate is raised in thyrotoxicosis and one sign of this is a high heart rate. The rate quoted is within the normal range.
   (b) **True** The body's food reserves become depleted if the rise in metabolic rate is not matched by an adequate increase in food intake.
   (c) **True** Triiodothyronine and thyroxine (tetraiodothyronine) are the thyroid hormones, and generally both are produced in excess in thyrotoxicosis.
   (d) **False** When the thyroid is producing excessive amounts of hormones, the level of TSH is depressed by a negative-feedback action.
   (e) **False** Increased appetite [see (b)].

83 (a) **False** The blood glucose was initially within normal fasting limits, and it returned to within the normal range soon after the injection.
   (b) **True** The low plasma pH and low bicarbonate are consistent with the production of ketoacids in diabetes mellitus. The PCO_2 is low because there has probably been some respiratory compensation for the non-respiratory acidosis.
   (c) **True** Ketoacids formed by catabolism of fats are excreted in the urine.
   (d) **True** The lack of insulin reduces glucose entry into the cells. Gluco-neogenesis therefore occurs and the consumption of fats and proteins leads to a loss in body weight.
   (e) **True** Blood sugar is raised above the renal threshold, and the excretion of glucose in the urine results in an osmotic diuresis. This can be sufficiently severe to necessitate emptying the bladder during the night (nocturia).

84 (a) **True** But only if it is charged and there is an electrical gradient in its favour.
   (b) **False**
   (c) **False** It is slowed down much less than this (about $2 \cdot 5\%$ per °C).
   (d) **True**
   (e) **False**

85 (a) **True**
   (b) **True** These are the properties required
   (c) **True** if a transport system is to be considered
   (d) **False** active.
   (e) **True**

86 The following are examples of active transport:
  (a) Chloride shift between red blood cells and the plasma.
  (b) Sodium reabsorption in the distal tubules of the kidney.
  (c) Movement of oxygen from pulmonary alveoli into the blood.
  (d) Uptake of calcium by the sarcoplasmic reticulum of muscle.
  (e) Oxygen movement within a muscle fibre.

87 Ultrafiltration is responsible for the fluid movement that takes place in the following processes:
  (a) Concentration of bile.
  (b) Salivation.
  (c) Glomerular filtration.
  (d) Sweating.
  (e) Interstitial fluid production.

N.B. Q88 is a searching question which has been included to test your understanding rather than as an example of an examination question.

*88 Suppose we have two compartments separated by a membrane (properties to be specified below). Compartment 1 is filled with a weak solution of NaCl and compartment 2 is filled with a solution of KCl at the same concentration.

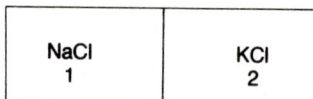

NaCl 1	KCl 2

  (a) If the membrane is equally permeable to Na^+, K^+ and Cl^-, redistribution of ions will occur until the two compartments have the same composition.
  (b) If the membrane is more permeable to K^+ and Cl^- than to Na^+, the Na^+ concentration in compartment 1 will be greater than that in compartment 2 when all net movement of ions ceases.
  (c) If the two compartments have rigid walls and the membrane is permeable to K^+ and Cl^- but not to Na^+, compartment 1 will be positive with respect to compartment 2 when all net movement of ions ceases.
  (d) If the two compartments have freely distensible walls and the membrane is permeable to K^+ and Cl^- but not to Na^+, compartment 1 will expand and compartment 2 will shrink.

86 (a) **False** The electrical gradient due to HCO_3^- loss from red cells favours movement of Cl^- into the cells.
   (b) **True**
   (c) **False**
   (d) **True** The uptake of calcium from sarcoplasm is against a concentration gradient.
   (e) **False** This is an example of facilitated diffusion, in which myoglobin is the carrier.

87 (a) **False** This is by active transport of Na^+ and Cl^- out of the gall bladder as an isosmotic solution.
   (b) **False** This is by active ion transport into the acinar lumen (probably Na^+) followed by isosmotic water transfer.
   (c) **True**
   (d) **False** Sweating is a similar process to salivation.
   (e) **True**

88 (a) **True**
   (b) **False** If the membrane is even slightly permeable to all three ions the compositions of the solutions in the two compartments will eventually be the same.
   (c) **True** K^+ will diffuse down its concentration gradient from compartment 2 to 1, setting up a voltage gradient across the membrane which causes Cl^- to follow. Eventually an equilibrium is reached when $[K^+]_1 \times [Cl^-]_1 = [K^+]_2 \times [Cl^-]_2$. The tendency for K^+ and Cl^- to move down their concentration gradients is then opposed by a voltage difference between the two compartments.
   (d) **True** The ion movements described in (c) will occur, but this results in osmotic imbalance between the two compartments. If the walls are distensible, water will therefore move from compartment 2 to compartment 1.

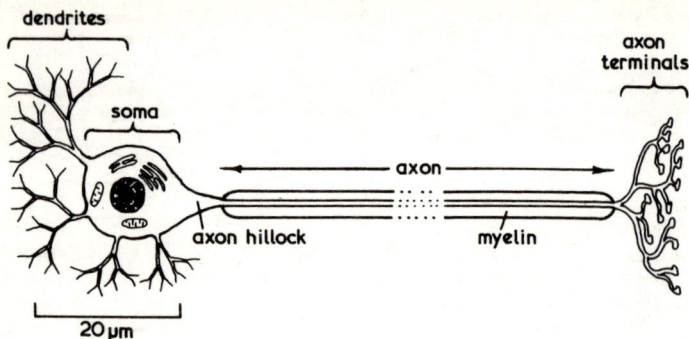

dendrites

soma

axon

axon terminals

axon hillock

myelin

20 μm

89 The diagram is of a motoneurone. The following statements refer to parts of the cell.
   (a) Macromolecules such as membrane constituents and the enzymes required to synthesize neurotransmitters are formed in the soma.
   (b) The initial segment has a lower threshold for firing an action potential than the rest of the soma membrane.
   (c) Suppose this motoneurone innervates the thumb of an adult human. The scale of the diagram is 2 cm bar = 20 μm. Drawn on this scale the full length of the axon would be about 1000 m (1 km).
   (d) There is a retrograde transport of chemicals along the axon, i.e. from the nerve terminals back to the soma.
   (e) Most of the inhibitory synapses are found on dendritic branches.

90 (a) There are about 10 times as many neurones as there are glial cells in the CNS.
   (b) Glial cells have a membrane potential that is close to the Nernst potential for potassium ions.
   (c) Oligodendroglia are found mainly in the grey matter of the CNS.
   (d) Protoplasmic astroglia can have end-feet on both blood vessels and on neurones.
   (e) Microglia are of different embryonic origin from astroglia.

89 (a) **True** They are then conveyed to the part of the neurone where they are needed.

    (b) **True** The lower threshold for the action potential at the initial segment than at the soma membrane means that the initial segment tends to act as the trigger zone for action potentials. The initial segment is an electrophysiological concept; it may be the same region as the axon hillock which is defined anatomically.

    (c) **True** The scale is × 1000. The true distance from the soma in the spinal cord to the thumb is about 1 m.

    (d) **True** Retrograde transport coexists with fast axonal transport and a slower flow of materials in the orthograde direction, from soma to nerve terminals.

    (e) **False** Inhibitory synapses are found mainly on the soma while most of the excitatory synapses are found on the dendrites.

90 (a) **False** There are about 10 times as many glia as neurones.

    (b) **True** The principal intracellular cation is potassium. The membrane potential is approximately 90 mV inside negative.

    (c) **False** The oligodendroglia are found mainly in the white matter, forming myelin around large axons, and surrounding bundles of small diameter axons.

    (d) **True** There are two types of astroglia: (1) fibrous astroglia which are found in the white matter, and (2) protoplasmic astroglia which are the satellite cells in the grey maner, forming a network between blood vessels and neurones. Their functional role is still uncertain.

    (e) **True** They are of mesodermal origin, and are homologous with the reticulo-endothelial cells in the body. Microglia are capable of migration and phagocytosis.

91 (a) Myelin is a poor electrical conductor compared with extracellular fluid.
   (b) In myelinated axons, the inward Na$^+$ current flow associated with an action potential only occurs at the nodes of Ranvier.
   (c) Myelin effectively increases the capacity of the nerve membrane.
   (d) Several unmyelinated C fibres are surrounded by a single Schwann cell.
  *(e) Following injury to a nerve that causes degeneration of the axon peripheral to the injury, the myelin sheath also disintegrates along the peripheral part of the nerve.

92 The left-hand diagram shows a frog sciatic nerve lying across a number of electrodes. A and B are used for stimulating and C and D for recording. The right hand diagram shows a typical recorded action potential.

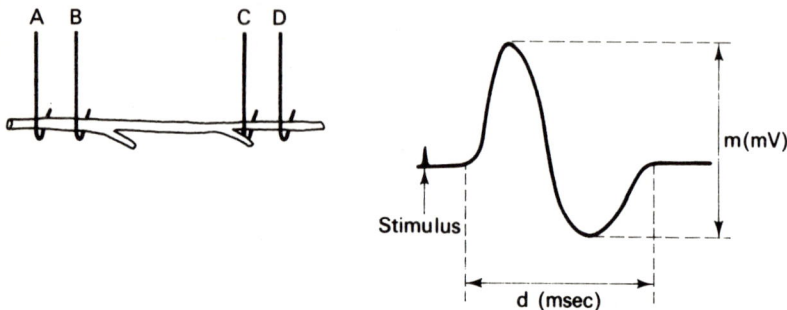

   (a) The first deflection on the recording occurs when electrode C is negative with respect to D.
   (b) The magnitude of the recorded action potential (m) will be independent of the distance between electrodes C and D.
   (c) The duration of the recorded action potential (d) will be independent of the distance between electrodes C and D.
   (d) The duration of the recorded action potential will depend on the distance between B and C.
   (e) The recorded action potential can be made monophasic by crushing the nerve (or applying local anaesthetic) at A.

91 (a) **True** The main current flow across the axon membrane occurs at nodes of Ranvier, where there are gaps between the Schwann cells that form the myelin sheath. Nodes occur every one to two millimetres along the axon.

(b) **True** Regenerative action potentials are localized to the nodes, and impulse conduction is saltatory (it jumps from node to node).

(c) **False** The decreased capacity of the nerve membrane increases the conduction velocity of the nerve impulse, because less current is needed to charge the membrane.

(d) **True** Many unmyelinated axons lie within invaginations of the Schwann cell membrane.

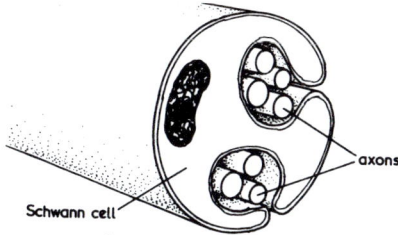

(e) **True** However, Schwann cells guide axon sprouts from the regenerating nerves to grow in the right direction. After the nerve has regenerated it is remyelinated.

92 (a) **True** The first deflection occurs when the wave of depolarisation, travelling from B along the nerve, reaches electrode C.

(b) **False** If the electrodes are close together, depolarization may reach D at a time when there is still depolarization under C. There would then be a smaller potential difference between the electrodes than if D were at a more distant site.

(c) **False** You should be able to work this out from the explanation given in (b).

(d) **True** The further apart the stimulating and recording electrodes, the more dispersed are the action potentials in fast and slow conducting fibres when they reach C, and hence the longer the wave duration.

(e) **False** By crushing or local anaesthetic at D.

93 The following question concerns the resting potential of a nerve fibre.
   (a) There is a higher concentration of K^+ ions outside the nerve than inside.
   (b) The potential across an axon membrane is largely determined by the logarithm of the ratio of the concentrations of K^+ ions outside:inside.
   (c) The membrane potential is positive inside relative to the outside.
   (d) The concentration ratio of Na^+ ions across the membrane contributes to the membrane potential to an extent governed by the permeability of the membrane to Na^+.
   (e) In the absence of a Na^+-K^+ pump, Na^+ entry would gradually depolarize the nerve.

94 The statements in this and the following question are related to the diagram below.

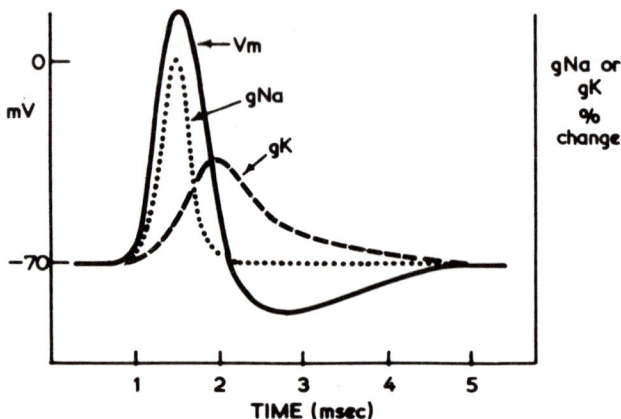

TIME (msec)

The graph shows a record of the voltage changes in an action potential recorded from an axon with an intracellular microelectrode (solid line: Vm).
The superimposed dotted and dashed lines show the *reconstruction* of the time courses of the changes in Na and K conductances (gNa and gK) respectively, obtained from voltage clamp experiments.
   (a) The conductance of an ion is directly proportional to the resistance of the membrane across which the ion moves.
   (b) Prior to the action potential the conductances of both K and Na are zero.
   (c) The net flow of Na current at the start of the action potential is inward.
   (d) The delay in the increase of gK after the onset of the increase in gNa enables the membrane to depolarize.
   (e) There is also an increase in the Ca^{2+} conductance during the action potential.

95 Acetylcholine is the chemical transmitter at:
   (a) All neuromuscular junctions in the somatic nervous system.
   (b) All post-ganglionic sympathetic endings.
   (c) All autonomic ganglia.
   (d) All post-ganglionic parasympathetic effector endings.
   (e) Any sites that are blocked by atropine.

93 (a) **False** Extracellular fluid contains about 5 mmol/l K^+ in mammals, while the intracellular K^+ concentration is about 150 mmol/l (or mmol/Kg of cell H_2O) in nerve.

(b) **True** The membrane potential V_m is close to E_K and
$$E_K = RT/F \log[Kout]/[Kin]$$

(c) **False** The excess concentration of K^+ inside the axon results in the K^+ ions tending to diffuse towards the outside of the membrane. The resulting separation of charge between them and the impermeant anions makes the inside of the membrane negative to the outside.

(d) **True** The permeability of the resting axon membrane to Na^+ is very much less than to K^+, so it produces only a small depolarization away from E_K.

(e) **True** The pump is essential for removing the Na^+ that leaks across the membrane and the Na^+ that enters in the action potential. However, these are small quantities, so it may take several minutes after poisoning the pump to detect a reduction in the membrane potential.

94 (a) **False** It is inversely proportional, $g = 1/R$, or $g = I/V$ where g is conductance, V voltage and I current.

(b) **False** It is the % *change* in gNa and gK that is plotted, so you cannot tell from the diagram. In fact there is a low gK and an even lower gNa at rest.

(c) **True** Na enters the nerve since there is a favourable concentration gradient and a favourable voltage gradient.

(d) **True** If the K ions moved out simultaneously with the inward movement of Na ions there would be no net flow of current across the membrane.

(e) **True** The entry of Ca^{2+} in the action potential is important at the axon terminals for transmitter release.

95 (a) **True**

(b) **False** Noradrenaline is the transmitter at most of these endings.

(c) **True**

(d) **True**

(e) **True** Atropine blocks the effects of acetylcholine released at postganglionic parasympathetic junctions (muscarinic effects of acetylcholine).

96 The following are statements about the role of calcium in nerve and muscle:
  (a) Calcium provides a link between excitation and contraction in all types of muscle.
  (b) The Ca^{2+} concentration is much lower in the cytoplasm of nerve and muscle fibres than it is in the extracellular fluid.
  (c) A reduction in the extracellular Ca^{2+} concentration decreases the excitability of nerve and muscle fibres.
  (d) An increase in the extracellular Ca^{2+} concentration results in an increase in the strength of contraction of skeletal muscle fibres.
  (e) More transmitter is released from nerve terminals by an action potential when the extracellular Ca^{2+} concentration is increased.

97 The following are statements about the properties of cardiac muscle:
  (a) Action potentials propagate from one muscle fibre to another.
  (b) The action potential lasts almost as long as the mechanical response.
  (c) Normal heart beats are neurogenic in origin.
  (d) A fused tetanic response can be produced by repetitive stimulation.
  (e) Spontaneous contractions occur in the muscle due to the presence of pacemaker cells.

98 The following statements are about motoneurones and skeletal muscle.
  (a) A single motoneurone invariably innervates more than one muscle fibre.
  (b) A motor unit in a postural muscle may have 100 muscle fibres.
  (c) A single impulse in the motoneurone gives rise to an action potential in each of the muscle fibres it innervates.
  (d) Modulation of motoneuronal firing frequency is the only way of changing tension in a muscle.
  *(e) When the $[Ca^{2+}]$ in the extracellular fluid falls drastically, the excitability of nerve and muscle is reduced.

96 (a) **True**
   (b) **True** It is actively taken up by sarcoplasmic reticulum in muscle, and extruded from nerve.
   (c) **False** Ca^{2+} stabilizes the membrane so a fall in extracellular Ca^{2+} increases excitability. At concentrations below normal, but higher than in overt tetany, the facial nerve can be activated merely by tapping the overlying skin (a sign of latent tetany).
   (d) **False** Ca^{2+} required for excitation–contraction coupling in skeletal muscle is provided by internal stores and Ca^{2+} is recycled within the cell.
   (e) **True**

97 (a) **True**
   (b) **True**
   (c) **False** They are myogenic, arising in pacemaker tissue. The autonomic nerve supply can affect the *rate* of firing in pacemaker cells.
   (d) **False** The long action potential means that the cell membrane is refractory throughout most of the mechanical response.
   (e) **True**

98 (a) **True** A single motoneurone and the group of muscle fibres that it innervates is called a motor unit.
   (b) **True** However, in muscle fibres where fine motor control is required e.g. extra-ocular muscles, there may be as few as three muscle fibres in each motor unit.
   (c) **True** Whether the motor unit contracts or not is therefore determined by the firing of the motoneurone.
   (d) **False** In a whole muscle, a graded increase in tension is produced by recruiting more motor units as well as by increasing the firing rate of the motoneurones of each motor unit. The tension developed in a muscle depends also on muscle length.
   (e) **False** Nerve and muscle membranes become hyperexcitable, and fatal spasm of the larynx can ensue. The condition is called tetany. A fall in Mg^{2+} in the extracellular fluid also increases excitability, since Mg^{2+} has a synergistic action to Ca^{2+} in stabilizing membranes. The action of divalent cations on the membrane is separate from their action on transmitter release.

99 These statements refer to the mechanisms of excitatory chemical transmission that have been observed in spinal motoneurones:

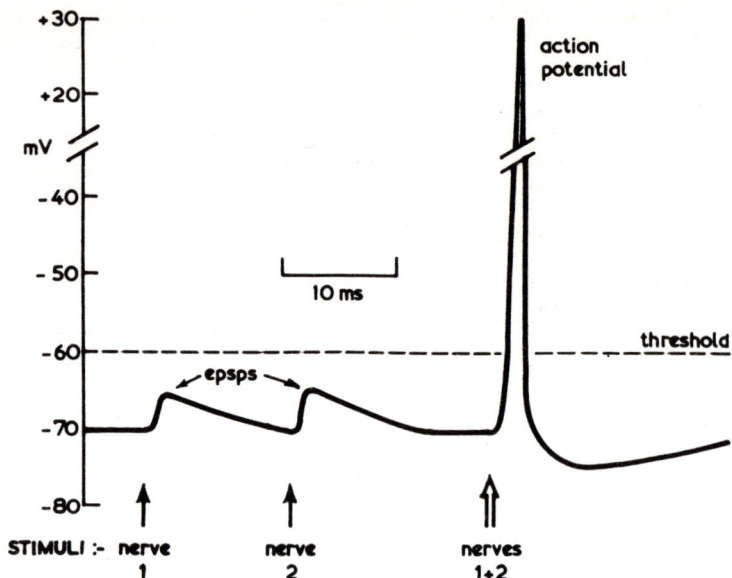

(a) An excitatory postsynaptic potential (epsp) produced in a motoneurone by stimulating muscle afferents is a graded process capable of summation.
(b) The threshold of excitation of a motoneurone refers to the minimum size of an epsp that is necessary to produce an action potential.
(c) Spatial summation refers to the additive nature of postsynaptic responses arising from separate afferent inputs.
(d) Temporal summation refers to the mechanism that causes repetitive firing of motoneurones.
(e) If the recording in the figure is from a motoneurone of an extensor muscle, then nerves 1 + 2 come from flexor muscles, acting at the same joint.

100 The following statements refer to mechanisms which have been shown to be involved in inhibition of neurones at some sites in the CNS.
(a) Release of the chemical transmitter glycine.
(b) Release of the chemical transmitter γ-amino butyric acid (GABA).
(c) Non-specific increase in conductance of the postsynaptic membrane to anions and cations.
(d) An inhibitory postsynaptic potential (ipsp) may hyperpolarize the membrane.
(e) During an ipsp a neurone cannot discharge a spike (action potential).

99 (a) **True** The epsp is a function of the size of the incoming volley. The epsps elicited by volleys in separate nerves may add together (see diagram).

(b) **False** Threshold (see diagram) refers to the membrane potential of the motoneurone that must be exceeded in order to excite an action potential. The threshold may vary with time according to the degree of activity of the motoneurone.

(c) **True** See response to stimulation of nerves 1 + 2 in diagram, which reached firing threshold.

(d) **False** Temporal summation is the summation of postsynaptic potentials (psps) that occurs when the afferent inputs are separated by time intervals shorter than the duration of their postsynaptic responses.

(e) **False** The epsps in an extensor motoneurone would arise from extensor muscle nerve stimulation. Flexor muscle nerve stimulation would produce inhibitory postsynaptic potentials (ipsps).

100 (a) **True** For example, glycine mediates postsynaptic inhibition in the spinal cord.

(b) **True** GABA has been implicated in presynaptic as well as postsynaptic inhibition.

(c) **False** There are specific increases to Cl^- or K^+. Non-specific increase in conductance of Na^+, K^+ and Cl^-, which occurs at the neuromuscular junction results in depolarization leading to excitation of the post-synaptic cell.

(d) **True** However, little or no change in membrane potential would take place if it was already close to the equilibrium potential of the ion which has had its conductance increased.

(e) **False** Excitatory and inhibitory potentials tend to summate algebraically so that a preponderance of epsps over ipsps may exceed threshold for spike initiation.

101 This question concerns peripheral and central neurotransmitters and neuro-
     modulators.
     (a) Although 5-hydroxytryptamine (5HT) is a neurotransmitter in the gut, it is not
         found in the central nervous system.
     (b) Noradrenaline is not found in neurones in the central nervous system.
     (c) Glutamate is an inhibitory transmitter.
     (d) The peptide substance P, is found in peripheral motor nerve axons.
     (e) Somatostatin, vasointestinal peptide and cholecystokinin are peptides which
         are found both in the gut and in neurones of the brain.

102 The following are statements concerning spinal cord reflexes:
     (a) The flexion reflex is elicited by stimulation of muscle spindles.
     (b) The tendon jerk or stretch reflex has a monosynaptic component.
     (c) Reciprocal inhibition of antagonist muscles of the same limb occurs during
         the stretch reflex.
     (d) Reflex withdrawal of a leg is usually accompanied by excitation of contra-
         lateral extensor motoneurones.
     (e) Golgi tendon organs are responsible for eliciting the clasp-knife reflex.

*103 The following statements refer to the consequences in man of a complete spinal
     cord section at a low thoracic (T10) site.
     (a) Permanent paralysis of voluntary movement of the legs.
     (b) Reflexes in the legs are enhanced for the first few days.
     (c) Muscle tone gradually develops in leg flexor muscles.
     (d) A few weeks after spinal cord section, stroking the outer border of the sole
         of the foot will produce plantar flexion of the big toe.
     (e) Spinal shock is the result of local damage to nervous tissue at the site of
         the lesion.

101 (a) **False** 5-HT (called serotonin in the USA) is found for example in neurones of the mid-brain, raphe nucleus, and hypothalamus as well as in the gut.

 (b) **False** Noradrenaline containing neurones are found in the medulla and pons, with the majority in the locus coeruleus of the pons.

 (c) **False** Glutamate is excitatory.

 (d) **False** It is found in some sensory nerves, innervating the tooth pulp and skin. It may be involved in the skin vasodilator response to skin crush or heat.

 (e) **True** Their function in the brain is not clear.

102 (a) **False** The flexion reflex is a protective reflex. Afferents excited by noxious stimuli activate interneurones which, in turn excite several flexor motor unit pools.

 (b) **True** Spindle afferents have been shown in the cat to synapse directly with motoneurones to the same muscle. Latency studies suggest that the reflex is also monosynaptic in man. A persistent response to stretch (the tonic stretch reflex) can occur in certain pathological states but it uses polysynaptic pathways.

 (c) **True** An interneurone link is involved changing the sign of action from excitation to inhibition.

 (d) **True** During ipsilateral flexion, in response to a noxious stimulus, a complementary extension of the contralateral leg readjusts support for the body.

 (e) **False** The reflex is elicited by strong muscle contraction that excites high threshold mechanoreceptors innervated by small diameter axons.

103 (a) **True** The lesion divides the descending motor tracts. The cut axons do not regenerate.

 (b) **False** Despite the fact that reflex arcs are still intact, all reflexes (deep and superficial) are depressed or absent immediately after cord section. The depression is known as spinal shock. Spinal shock gradually lessens over several weeks.

 (c) **True** The legs may even adopt a posture known as paraplegia in flexion with the hip, knee and ankle slightly flexed.

 (d) **False** The diagram shows the normal adult response of plantar flexion (a) and the Babinski sign of dorsiflexion (b) which is an abnormal response found after transection of the corticospinal tract. The Babinski sign is normal in infants because the corticospinal tract is not fully developed.

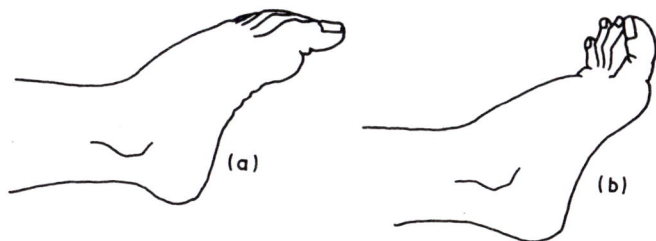

 (e) **False** Shock is due to interruption of descending spinal tracts. After recovery, a second lesion below the original cut does not reintroduce spinal shock.

RIGHT　·　LEFT　　　R　　　L　　　R　　　L
DORSAL

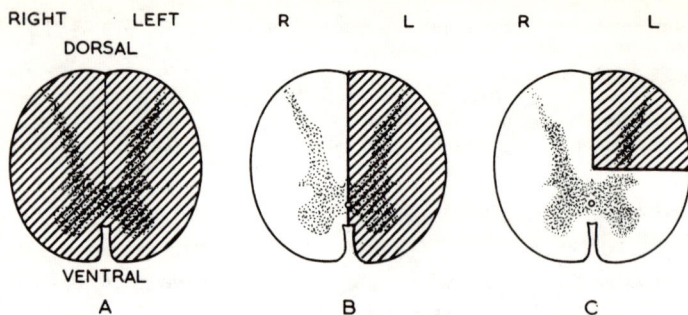

VENTRAL
　　　A　　　　　　　　B　　　　　　　　C

*104 The question refers to the diagrams A to C, which show the position of three different lesions (hatched lines) in the spinal cord at T8.
(a) Following lesion B, there will be paralysis of voluntary movement of the right leg.
(b) An exaggerated extensor tone will gradually appear two to three weeks after lesion A.
(c) After lesion C, there will be a loss of pain and temperature sensation from dermatomes below T8 on the right half of the body.
(d) In the days immediately following lesion A there will be retention of urine in the bladder.
(e) A few weeks after lesion A the lower half of the body will be hot and dry to the touch.

*105 This question is about the light reflex.
(a) When a bright light is shone into one eye, the pupils of both eyes constrict.
(b) The afferent pathway for the pupillary light reflex passes from the retina to the lateral geniculate nucleus.
(c) Damage to the oculomotor nerve could lead to failure of the pupillary response to light.
(d) Contraction of the sphincter muscle of the iris is brought about by the noradrenergic sympathetic innervation.
(e) During fixation onto a near object (when accommodation and convergence of the eyes occur) the pupils also constrict.

104 (a) **False** The pyramidal and extrapyramidal pathways necessary for the execution of voluntary movements have crossed before entering the spinal cord.

   (b) **False** It is an exaggerated flexor tone that develops [see answer 103 (c)].

   (c) **False** Since the fibres subserving pain and temperature sensation run in the ventrolateral quadrant of the cord, the fibres from both sides of the body would still be intact.

   (d) **True** After a few weeks the bladder empties reflexly when it is full.

   (e) **False** The immediate effects of the lesion include loss of vasomotor and sudomotor tone, but after recovery from spinal shock, the lower half of the body will tend to be cold and sweaty.

105 (a) **True** The constriction of the pupil in the same eye is called the direct light reflex, and that of the other eye, the consensual light reflex.

   (b) **False** It goes to the pretectum an area just rostral to the superior colliculus. The axons from the pretectal nuclei terminate in the Edinger–Westphal nuclei.

   (c) **True** The efferent limb of the reflex involves autonomic preganglionic axons, which run from the Edinger–Westphal (accessory oculomotor) nucleus via the oculomotor (IIIrd cranial) nerve to the ciliary ganglion.

   (d) **False** It is brought about by cholinergic parasympathetic fibres, which originate in the ciliary ganglion. Noradrenergic sympathetic activity produces dilatation of the pupil by contracting the radial muscle of the iris.

   (e) **True** The constriction of the pupil reduces light scattering in the eye and improves visual acuity.

*106 Questions 106 and 107 concern some of the functions of the cranial nerves.

  (a) A patient has double vision (diplopia) and her right eye is turned inwards. A possible cause is damage to the right abducens (VIth cranial) nerve.

  (b) Paralysis of the upper eyelid can be caused by damage to the oculomotor (IIIrd cranial) nerve.

  (c) A drooping left eyelid, seen together with a constricted left pupil and flushing of the left side of the face, is likely to be caused by damage to the left oculomotor nerve fibres.

  (d) Inability to move the right eye downwards and towards the nose can be the result of damage to the innervation of the right inferior oblique eye muscle.

  (e) Failure to move the eyes together (conjugate deviation) to the right can be caused by a lesion in the pons on the right side of the brain.

106 (a) **True** The abducens nerve innervates the lateral rectus muscle. The eye deviates inwards horizontally due to the unopposed action of the medial rectus muscle.

(b) **True** The oculomotor nerve innervates the striated levator palpebrae muscles. Among other signs, there may be dilatation of the pupil if the parasympathetic fibres in the IIIrd nerve are damaged.

(c) **False** This collection of signs (known as Horner's syndrome) can be produced by damage to the fibres of the superior cervical (sympathetic) ganglion. Sympathetic nerve activity contracts the smooth muscle of the upper eyelid, dilates the pupil and causes vasoconstriction.

(d) **False** The superior oblique muscle, innervated by the trochlear (IVth cranial) nerve is responsible for depressing the eye. A unilateral trochlear nerve lesion can also give rise to a vertical diplopia.

(e) **True** See the diagram (below) of pathways and cranial nerves involved. Another cause could be a left sub-cortical lesion.

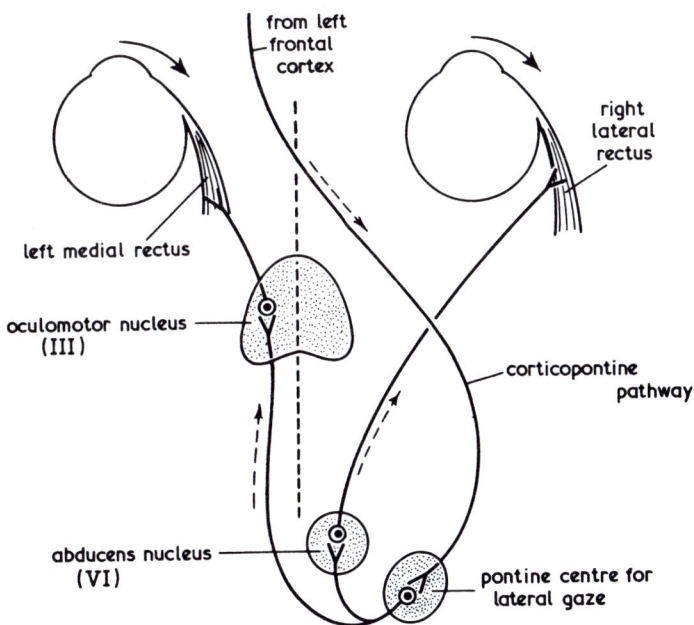

Pathways concerned in conjugate horizontal movement to the right

*107 (a) A patient is asked to stick out his tongue to the front. If the left side of the tongue shows muscle wasting, and it deviates to the left, this would suggest damage to the right hypoglossal (XIIth cranial) nerve.

(b) A complete lesion of the right facial (VIIth cranial) nerve causes more widespread paralysis of facial muscles than does a lesion affecting all the fibres projecting from the left motor cortex to the face.

(c) The gag reflex (contraction of the pharynx elicited by touching its walls) may be absent following damage to the glossopharyngeal (IXth cranial) nerve.

(d) All the motor fibres of the vagus (Xth cranial) nerve are part of the autonomic nervous system, innervating glands, smooth and cardiac muscle.

(e) Weakness in turning the head to the right and shrugging the left shoulder could be due to damage to the left accessory (XIth cranial) nerve.

108 Sensory receptors transduce one form of energy into another. The final form of this transformation is the action potential which propagates into the central nervous system. Excitation of peripheral receptors does not automatically lead to the generation of sensation. Bear this in mind when assessing the following statements:

(a) Individual peripheral receptors are sensitive to only one particular form of energy within their normal physiological range of working.

(b) Adapatation of a receptor to a constant stimulus can be a function of the receptor structure rather than the nerve terminal.

(c) Touch, vibration, temperature and pain are classed as different modalities of cutaneous sensation.

(d) Quality of sensation refers to the recognizable subdivisions within a certain modality of sensation.

(e) The increase in frequency of discharge in an afferent axon is the only way in which an increase in the strength of a stimulus can be signalled to the CNS.

109 The following statements are concerned with somatic and visceral pain.

(a) Intense stimulation of any peripheral nerve fibres or receptor types results in pain.

(b) Pain can be divided into two components, a well localized, pricking pain which is rapidly conducted, and a poorly located, slowly conducted pain sensation which outlasts the cause or stimulus.

*(c) In the Brown-Séquard syndrome, which results from unilateral damage to the spinal cord, patients report a loss of pain and thermal sensation ipsilaterally to the lesion and below the level of section.

(d) The hollow organs of the viscera are insensitive to touch, temperature change and cutting.

(e) Pain is the only sensation to be elicited by distension of the hollow organs.

107 (a) **False** The left hypoglossal nerve, since muscle wasting occurs after motor nerve damage. The deviation of the tongue is to the damaged side as a result of normal muscle activity on the right side.

 (b) **True** This is because there is a bilateral representation of the facial muscles in each primary motor cortex. The unaffected cortical hemisphere can still control upper facial movements.

 (c) **True** Sensory fibres from the pharynx travel in the glossopharyngeal nerve. Motor fibres from the nucleus ambiguus in the medulla and pons travel to the pharynx and soft palate in the IXth and Xth nerves.

 (d) **False** Most of the efferents are autonomic, but there are also vagal motor fibres to the striated muscles of the soft palate, pharynx and larynx. They are involved in swallowing, coughing and clearing the voice.

 (e) **True** The nerves innervating the trapezius and sternocleidomastoid muscles have cell bodies in C1 to C5, and comprise the spinal part of the accessory cranial nerve.

108 (a) **False** Some receptors, which are innervated by slowly conducting myelinated (A delta) or unmyelinated (C) fibres, are termed polymodal receptors since they respond to several (thermal, mechanical, chemical) energy forms.

 (b) **True** For example, removing the lamellae from a Pacinian corpuscle to expose the bare nerve terminal changes the receptor response from a rapidly adapting to a non- or slowly adapting one.

 (c) **True** It is usual to find different receptor types subserving the different modalities. In the skin, for example, there are free nerve endings for pain and encapsulated endings for touch.

 (d) **True** Examples of quality of sensation are colours in vision, and the different qualities of taste (sweet, sour, bitter, salt).

 (e) **False** Spatial coding (number of channels) is also used. An increasing strength of stimulation excites other afferents as well as raising the discharge frequency of those already recruited. Rarely, receptors may use just the one code. Rapidly adapting Pacinian corpuscles are an example of a population of receptors which may utilize only the spatial code.

109 (a) **False** This is an outmoded concept. There are specific nerve endings which are nociceptive, i.e. they respond to noxious or harmful stimuli.

 (b) **True** The sharp pain is conveyed in small, myelinated (A delta) axons whereas the more diffuse pain travels in more slowly conducting, unmyelinated (C) fibres.

 (c) **False** The ascending spinothalamic fibres originating from cells in the dorsal horn decussate (cross over) in the spinal cord at their segmental level of origin. The loss of pain and temperature sensations is contralateral.

 (d) **True** One advantage of this lack of sensitivity is that surgery may be carried out using only local anaesthetization for the initial abdominal incision.

 (e) **False** Fullness and satiation may also be experienced.

110 These statements concern peripheral and central factors that determine the appreciation and intensity of pain.
   (a) Nerve section leads to a permanent loss of pain sensation.
   (b) Referred pain is the term given to the painful sensations that are felt in a peripheral structure, such as the arm, but which are associated with trauma to the deep viscera, such as the heart.
   (c) Lesions to the thalamus can produce raised thresholds for pain.
   (d) Lesions of the frontal lobes of the cerebral cortex alter a person's attitude to pain rather than the intensity of the pain.
   (e) The groups of naturally occurring peptides known as endorphins and enkephalins influence pain transmission in the CNS.

111 The following statements are about the optical system of the eye:
   (a) Most of the refractive power of the eye lies at the air/corneal junction.
   (b) In myopia (short-sightedness) the eyeball is too short for good visual acuity over the normal range of object distances from the eye.
   (c) Spectacles with biconcave lenses will correct defects of hypermetropia (long-sightedness).
   (d) Spherical lenses cannot fully correct astigmatism.
   (e) Visual acuity is better for central vision than for peripheral vision.

110 (a) **False** Temporary relief may be complete but, since the cell bodies of the primary neurones are intact, axons tend to regenerate. Phantom limb pain (as well as other phantom sensations) is thought to arise mainly from regenerating axons.

(b) **True** A characteristic of referred pain is that both the superficial and visceral structures are innervated by the same spinal roots.

(c) **True** The threshold for pain, as well as for touch, temperature and pressure, is raised on the contralateral side. However, sensory stimulation may now produce exaggerated responses and may lead to intractable pain. The condition is known as thalamic syndrome and suggests more than just a simple relay function for the thalamus.

(d) **True** Emotional responses are often detached from chronic pain following frontal lobotomy. A patient may admit that pain is as intense following the lesion but that it does not bother him to the same extent.

(e) **True** These peptides bind with high affinity to receptors in the CNS which also bind opiate drugs. Significantly, enkephalins are associated with interneurones in the dorsal horn of the spinal cord at the site of the first central connections of peripheral pain afferents.

111 (a) **True** About 2/3 of the refractive power of the eye lies at the air-corneal junction. The importance of the lens lies in the fact that its curvature, and hence its focal length, can be varied.

(b) **False** In myopia the eyeball is too long, so the images of all but the closest objects tend to be in focus in front of the retina.

(c) **False** In hypermetropia the eyeball is too short. Biconvex lenses help to focus images on the retina.

(d) **True** There is a different curvature of the refracting system in one plane than in another. Cylindrical lenses are used to improve astigmatism.

(e) **True** This is partly because the packing density of cones is greatest in the fovea. An additional factor is that there is less convergence in the neural pathway from the fovea than from the peripheral retina.

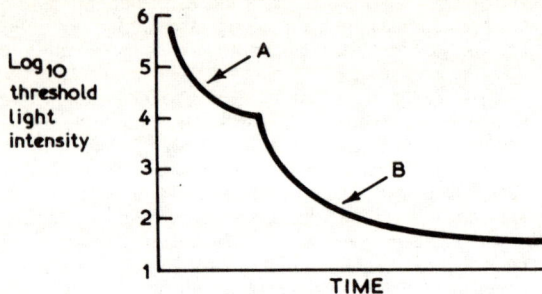

112 The graph shows dark adaptation curves for a human subject going from sun-
    shine into a dark room.
    (a) Curve A shows the time course of regeneration of the pigment rhodopsin.
    (b) The subject is unlikely to have prolonged vitamin A deficiency.
    (c) Dilation of the pupil makes the major contribution to curve A.
    (d) The light intensity required for vision is lower in the periphery of the retina
        than at the fovea.
    (e) The cones are responsible for colour vision.

113 (a) There are three main visual pigments in the normal human retina.
    (b) Colour blind dichromats can match all the hues they see with a mixture of
        two spectral colours.
    (c) Colour blindness is found more commonly in females than males.
    (d) The convergence of several cones onto a cone-bipolar cell reduces visual
        acuity.
    (e) Lateral inhibition in the retina increases contrast in the visual field.

114 This question is about relay neurones in the lateral geniculate nucleus (LGN) and
    neurones in the visual receiving areas of the cerebral cortex.
    (a) The majority of LGN cells can be excited from either eye.
    (b) LGN neurones project to the striate cortex on the same side of the brain.
    (c) Most neurones in striate cortex can be excited from both eyes.
    (d) There is more than one 'map' of the retina in the visual cortex.
    (e) The topography of the retina is mapped on the striate cortex with its propor-
        tions accurately preserved.

112 (a) **False** Rod vision is ineffective at high light intensities. Curve A shows dark adaptation for cones, and curve B for rods.

(b) **True** Vitamin A is a precursor of retinal, which is used to form rhodopsin. Consequently, a prolonged dietary deficiency of Vitamin A can cause night blindness, and the graph of dark adaptation would lack curve B.

(c) **False** It is a small effect as pupil area increases by only about four times.

(d) **True** Rods are absent from the fovea. A faint star can be seen more easily if you fixate to one side of it.

(e) **True** The rods are important for twilight vision.

113 (a) **False** There are at least four: the rod pigment rhodopsin and three pigments for colour vision. Erythrolabe, chlorolabe and cyanolabe absorb maximally the wavelengths of red, green and blue light respectively i.e. the three spectral colours.

(b) **True** Trichromats with normal colour vision need three spectral colours.

(c) **False** Many disturbances of colour vision are inherited with the X chromosome as recessive traits, and hence are commoner in males.

(d) **True** Cone acuity is greatest at the fovea where convergence is least.

(e) **True** Lateral inhibition enhances edge detection.

114 (a) **False** Although there are inputs to the LGN from each retina, they end in different cell layers in the nucleus, and remain segregated.

(b) **True** Axons from all layers of the left LGN project to the left striate cortex, and those from the right LGN to the right striate cortex.

(c) **True** The striate neurones are usually excited more strongly from one eye than the other. The axons of geniculo–cortical neurones excited by the left eye terminate mainly in vertical columns, or slabs, adjacent to those excited by the right eye.

(d) **True** There are separate maps of the retina in V1, V2 and V3. V4 does not have a retinotopic organization.

(e) **False** The map of the retina in striate cortex is greatly distorted. Eighty to ninety % of V1 is devoted to the representation of the fovea.

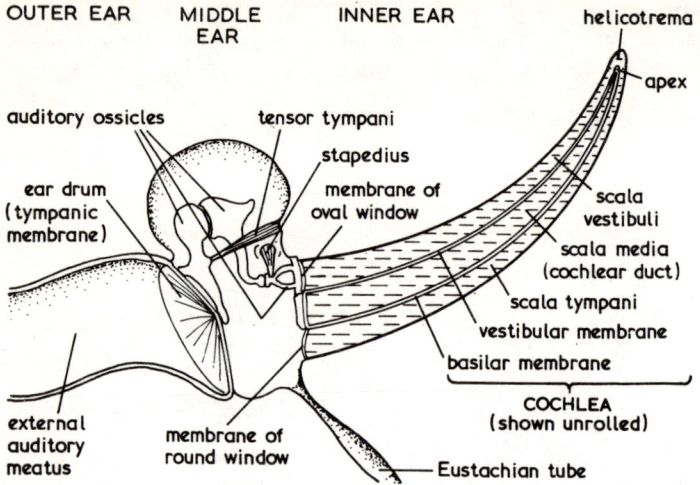

OUTER EAR   MIDDLE EAR   INNER EAR   helicotrema

apex

auditory ossicles   tensor tympani

stapedius

ear drum (tympanic membrane)   membrane of oval window   scala vestibuli

scala media (cochlear duct)

scala tympani

vestibular membrane

basilar membrane

COCHLEA (shown unrolled)

external auditory meatus   membrane of round window   Eustachian tube

Questions 115 to 118 concern the ear and hearing.

115 (a) Conduction deafness is invariably due to problems with conduction of sound along the ossicles of the middle ear.
  (b) The eustachian tube is normally open in the pharynx to allow equalization of pressure across the eardrum.
  (c) The function of the ossicles of the middle ear is to reduce the force/unit area at the oval window.
  (d) Contraction of the stapedius and tensor tympani muscles reduces the transmission of low frequency sounds to the oval window.
  (e) The oval window, but not the round window, vibrates when sound is conducted by the ossicles.

115 (a) **False** A reversible conduction deafness is found when wax blocks the outer ear or the tympanic membrane is inflamed.

(b) **False** It is usually closed at the pharyngeal end, but opens during swallowing, or yawning, or can be opened voluntarily. If it is blocked by inflamed, swollen membranes or secretions from the nose, equalization of pressure across the eardrum is prevented, and rupture of the eardrum could be produced by large pressure differences (diving, flying by aeroplane).

(c) **False** There is considerable gain in force/unit area at the oval window, with an associated reduction of amplitude of displacement, compared with that of the tympanic membrane. The ossicles form an impedance matching device.

(d) **True** A loud sound initiates the contraction reflexly (after about 40 msec). This attenuation reflex may have some protective function, or may mask low frequency background noise and thus enhance perception of higher frequencies.

(e) **False** Sound waves are conducted through the fluid of the inner ear from oval to round window. The latter bulges out into the middle ear when the oval window is displaced inwards, away from the middle ear.

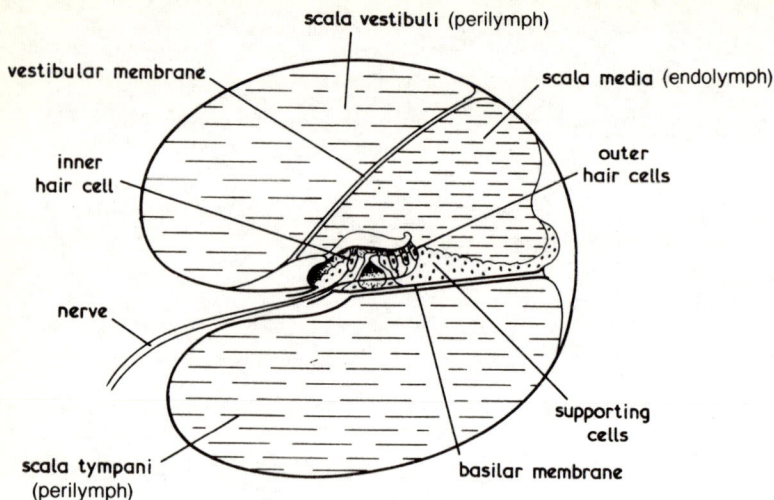

Cross section through cochlea

116 (a) The endolymph of the inner ear has Na$^+$ and K$^+$ concentrations similar to those of intracellular fluid.

(b) Displacement of the basilar membrane leads to bending of the cilia of the hair cells and the production of a receptor potential.

(c) The normal range of audible frequencies for humans is from 20 Hz to 50,000 Hz (50 kHz).

(d) For normal speech to be intelligible, we need to hear frequencies in the range 300 Hz to 3·5 kHz.

*(e) A patient reports normal hearing in his right ear, but impaired hearing in the left ear. If the stem of an oscillating tuning fork (e.g. at 256 Hz) is placed on the centre of the forehead (Weber's test) and the patient reports the sound is on the left side, the diagnosis would be impaired conduction in the left *outer* or *middle* ear.

116 (a) **True** The perilymph on the other hand is similar in composition to interstitial (extracellular) fluid.

(b) **True** The receptor potential is a potential difference between the endolymph in the scala media and perilymph, in the scala tympani. It is probably produced by metabolic pumping in cells of the stria vascularis.

(c) **False** The normal human range is from about 20 Hz to 16 kHz. Many animals (mice, owls, dogs and bats) can hear higher frequencies than humans.

(d) **True** From the average threshold curve for hearing (below), it can be seen that the ear is very sensitive over this range.

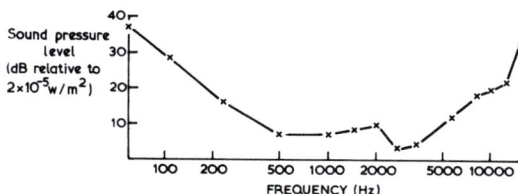

(e) **True** Inflammation of the middle ear, for example, can impair conduction from the eardrum to the round window. Bone conduction bypasses the normal route, exciting the cochlea directly. The sound appears louder on the side with impaired middle ear conduction. If the left inner ear had been affected, the sound would have been louder on the right.

117 This question concerns coding of frequency and sound intensity in the cochlear nerve.

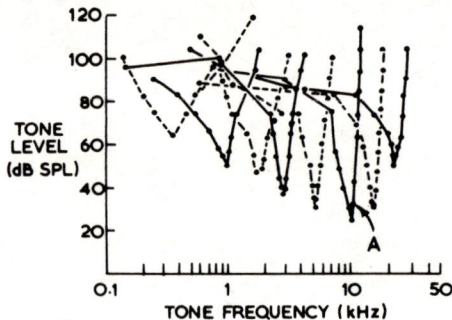

TONE LEVEL (dB SPL)

TONE FREQUENCY (kHz)

The graph shows an array of frequency threshold curves (i.e. the threshold for firing to single pure tones) for single cochlear nerve fibres in the guinea-pig.

(a) The fibre with curve A fires in response to a 10 kHz tone only when the intensity is about 60 dB greater than that of the 6 kHz tone.

(b) The cochlear nerve fibres whose threshold frequencies are shown in the graph above are likely to innervate receptors at different, systematically ordered, positions along the cochlea.

(c) It can be seen from the graph that as a tone gets more intense, more fibres in the cochlear nerve will fire.

(d) A greater displacement of the basilar membrane is perceived as a louder sound.

(e) There is a linear relationship between increase in the sound pressure and the perceived increase in loudness.

117 (a) **False** The peak sensitivity is at 10 kHz, and is called the characteristic
frequency of the fibre. The peak sensitivities of the whole array of fibres
follow closely the behavioural audiogram of the species.

(b) **True** The tonotopic organization of fibres in the nerve results primarily from
the mechanical characteristics of the basilar membrane.

(c) **True** Also the rate of firing in each fibre will tend to increase.

(d) **True** More hair cells are excited.

(e) **False** There is an approximately logarithmic relationship between increase of
sound pressure and each step of perceived increase in loudness.
Relative loudness is thus described well on the decibel (dB) scale.

118 This question is about the auditory cortex.
   (a) A difference in intensity of a sound at each ear is essential for localizing the source of a sound in space.
   (b) It is impossible to localize a sound in space when one ear is deaf.
   (c) A time delay between the arrival of a sound at each ear is one way in which the direction of a sound can be localized.
   (d) Neurones of the auditory relay nuclei in the pons, mid-brain and thalamus all receive excitatory inputs from both the ipsilateral and contralateral cochleas.
   (e) The auditory cortex is necessary for discrimination of the temporal patterning of sounds (i.e. tunes).

119 A subject with head tilted nose downward by about 30 degrees with reference to the normal upright position, is rotated in a chair toward the right, at 30 revs/min for one minute. The movement of the chair is then stopped. The rotation can be either with the eyes open (in the light) or with eyes closed (in the dark). The potential changes recorded between two electrodes placed 1 cm beyond the lateral edge of each eye are used to record eye movements. The questions relate to this experiment.

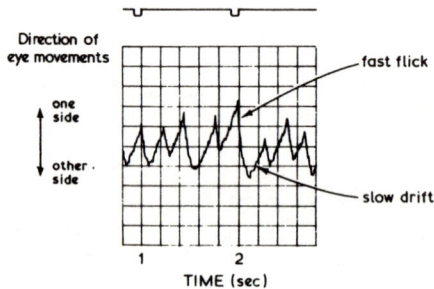

Record of nystagmus with eyes open, in the light.

   (a) Throughout the rotation with the eyes *open* you would expect to record nystagmus with a slow horizontal drift to the right and a fast flick to the left.
   (b) At the start of rotation with the eyes *shut*, you would expect to record nystagmus, with a slow horizontal drift to the right and a fast flick to the left.
   (c) After spinning at a constant angular velocity for about 30 seconds with the eyes *shut*, you would no longer expect to record nystagmus.
   (d) A subject showing no nystagmus when rotated in the dark could have suffered damage of the VIIIth nerve.
   (e) It is the stimulation of the otolith organs in the labyrinths that would produce the nystagmus.

118 (a) **False** It is information that can be used for localization, but other cues e.g. phase difference, may also be used.

(b) **False** It is harder to localize sounds, but the head can be moved to find the direction where the sound is loudest.

(c) **True** If clicks of equal intensity are presented to each ear, and one click is progressively delayed relative to the other by a few milliseconds, the click is still heard as a single sound, but the perceived origin of the sound moves around the head.

(d) **True** In the main visual and somatosensory afferent systems major convergence between inputs from each side of the body does not occur until cortical level. A lesion in the auditory cortex (temporal lobe) does not lead to deafness of one ear.

(e) **True** Also for localization of sounds in space.

119 (a) **False** The eye movements are produced by the attempt to fixate objects in the surroundings in spite of the head movement, so the drift is to the left (see tracing). The fast component is to catch up with the head movement, and is therefore to the right. The eye movements are called nystagmus, and the direction of nystagmus is given by the fast component. When there is relative movement between the visual fixation point and the head, the nystagmus is termed optokinetic nystagmus. The nystagmus continues throughout rotation with the eyes open.

(b) **False** Labyrinthine reflexes help to fixate the eyes in spite of the acceleration of the head. The labyrinthine nystagmus is therefore to the right, i.e. in the same direction as that produced by the visual input.

(c) **True** It is only rotational acceleration of the endolymph in the semicircular canals that initiates labyrinthine nystagmus. Once the endolymph is moving at the same velocity as the bony semicircular canals, the hair cells receive no stimulation.

(d) **True** Another cause can be damage to the hair cells, resulting for example from the toxic effects of aminoglycoside antibiotics such as streptomycin, gentomycin, kanamycin and auromycin.

(e) **False** The otoliths are the receptors in the saccule and utricle of the labyrinth, and are responsive to gravitational pull. Thus they give information about the static position of the head in space. The receptors for rotation are the cristae of the ampullae of the semicircular canals.

120 This question concerns the sense of taste and the sense of smell.
    (a) There are separate zones on the tongue with specific sensitivity for salt, bitter, sweet and sour tastes.
    (b) Other components of taste depend on activity in the olfactory (lst) cranial nerve.
    (c) A taste aversion can be produced by a single experience of a sapid substance (i.e. one with a taste) that produces unpleasant consequences.
    (d) A pheromone is an odorous externally secreted substance that can affect the behaviour or physiology of another individual of the same species.
    (e) Strong unpleasant smells activate the parasympathetic nervous system.

121 The ability to orientate the body in space makes use of afferent information from a variety of sources.
    (a) The otoliths (saccule and utricle) contain receptors which respond to gravitational force and linear acceleration.
    (b) The otoliths detect any misalignment of the head on the body.
    (c) Chronic loss of function of the labyrinths may be compensated if the person can see.
    (d) Proprioceptive afferent input from muscles, tendons and joints only contributes to our equilibrium when the body is moving.
    (e) Vision is necessary for righting reflexes.

*122 Disorders of motor performance often can be linked with damage or disease of a circumscribed part of the central nervous system.
    (a) Hypotonia is a classical sign of cerebellar disease.
    (b) Intention tremor is a sign of dorsal column lesion resulting from tertiary syphilis.
    (c) A pendular knee jerk is a sign of cerebellar damage.
    (d) Nystagmus is a sign of cerebellar damage.
    (e) Hemiballismus is a sign of cerebellar damage.

120 (a) **True**   The different zones are: bitter near the back of the tongue, sour and salty near the tip, and sweet along the edges.

(b) **True**   Olfactory stimuli contribute greatly to our sense of taste. A severe cold impairs taste sensation.

(c) **True**   Even when the unpleasant result (e.g. vomiting) is delayed for several hours, the substance is avoided for a long time thereafter. This one-trial learning has obvious survival value.

(d) **True**   For example, odours from mammalian (non-primate) urine have been shown to change female reproductive physiology (length of reproductive cycle, or implantation of fertilized ovum).

(e) **True**   They may produce lachrymation, salivation, bronchoconstriction, and nasal secretion. Pleasant smells associated with food provoke gastric secretions.

121 (a) **True**   They should not be confused with the semicircular canals, which are also in the bony labyrinths, but which respond to rotational accelerations.

(b) **False**   The misalignment is detected by muscle spindles and joint receptors in the neck.

(c) **True**   However, acute loss is accompanied by disequilibrium, nausea and nystagmus.

(d) **False**   It is a major source of afferent input used to maintain balance at rest as well as during movement.

(e) **False**   The labyrinthine and tonic neck reflexes are sufficient.

122 (a) **True**   Hypotonia is associated with damage to the flocculonodular lobes of the cerebellum. Afferent input to this region is largely vestibular.

(b) **False**   Intentional tremor is a characteristic of damage to the lateral lobes of the cerebellum (the neocerebellum). Tabes dorsalis, which is a consequence of tertiary syphilis, is characterized principally by peripheral sensory impairment. However, the loss of myelinated afferents can also give rise to hypotonia and so present difficulties in diagnosis.

(c) **True**   Stretch reflexes are impaired.

(d) **True**   Sustained fixation is difficult and the eyes drift toward the midline.

(e) **False**   It is seen following damage to the contralateral subthalamic nucleus.

123 These statements concern the physiology and disorders of the basal ganglia:
   (a) The caudate nucleus, the putamen and globus pallidus are sources of major motor tracts which descend to the spinal cord.
   (b) The most numerous and important connections of the basal ganglia, apart from their intrinsic connections, are with the cerebral cortex.
   (c) There is a marked sensory loss in basal ganglia disorders.
  *(d) Patients suffering from hemiballismus, due to a lesion of the sub-thalamus, have a profound loss of balance.
  *(e) High levels of the neurotransmitter, acetylcholine, are found in the caudate nucleus of the basal ganglia.

124 The cerebellum is part of the CNS which forms a particularly prominent structure in those vertebrates which have highly developed motor skills.
   (a) Projection of proprioceptive and exteroceptive impulses from the limbs to the cerebellum is predominantly ipsilateral.
   (b) Direct electrical stimulation of the cerebellum can elicit complex sensations.
   (c) Purkinje cells are the only neurones with efferent axons leaving the cerebellar cortex.
   (d) The synaptic connections of Purkinje cell axons are all inhibitory.
   (e) Direct electrical stimulation of the cerebellar cortex can give rise to discrete muscle contractions or movements.

125 This question is concerned with the major tracts that originate in particular areas of the brain and that descend the spinal cord to determine motor output:
   (a) The axons of the majority of pyramidal tract neurones synapse directly with spinal motoneurones.
   (b) Complete section of the pyramids in the medulla results in a permanent loss of precision in the performance of motor tasks.
   (c) The vestibulospinal tract to the lumbar spinal cord controls extensor rather than flexor motoneurone output.
   (d) Most tectospinal tract neurones control motoneurones that are used to achieve orientation of the head on the body.
   (e) The rubrospinal tract has its major influence on axial and girdle musculature that is used to maintain posture and equilibrium.

123 (a) **False** None does so. However, the red nucleus is included by some as a component of the basal ganglia and it is the origin of the descending rubro-spinal tract.

(b) **True** There is a projection from the cerebral cortex to the putamen and caudate nucleus and an efferent connection back to the cerebral cortex from the globus pallidus via the ventral thalamus.

(c) **False** The basal ganglia are involved in motor behaviour, particularly through their influence on the motor cortex.

(d) **False** In hemiballismus the extremities are thrown about violently, but the sufferer compensates for the movements and maintains remarkably good balance.

(e) **True** Other parts of the basal ganglia have noticeably high levels of monoamine transmitters (see question 127).

124 (a) **True** This is in contrast to projection to the cerebral cortex where sensory representation is contralateral.

(b) **False** No sensations are elicited. Again, this is in contrast to the result of stimulating the post-central gyrus of the cerebral cortex. The wealth of afferent input to the cerebellum thus implies a role other than the generation of sensation.

(c) **True** The vast majority of Purkinje cell axons terminate within the intracerebellar nuclei. The axons of these nuclear cells then leave the cerebellum. Purkinje cells also have collaterals which terminate within the cerebellar cortex.

(d) **True** They inhibit neurones in the deep cerebellar nuclei and cerebellar cortex.

(e) **False** There are no motor tracts originating in the cerebellar cortex that are analogous to the corticospinal or other descending tracts from the midbrain. Stimulation of the cerebellar nuclei can give rise to movements since they, in turn, influence the output of descending tracts such as the vestibulospinal.

125 (a) **False** Most synapse with interneurones. A number do synapse directly with motoneurones in the contralateral cord but these are largely restricted to distal muscles of the upper limbs in primates.

(b) **True** Pyramidal tract lesions affect motor skills involving intricate hand and finger movements. Locomotion, posture and the power needed to grasp are largely unaffected.

(c) **True** Extensor muscles of the legs are the antigravity muscles used in posture and are under the control of the vestibular system.

(d) **True** The tectum (superior and inferior colliculi) has important visual and auditory inputs. The tectospinal tract projects largely to the cervical spinal cord where it influences neck motoneurones.

(e) **False** The rubrospinal tract descends in the lateral funiculus close to corticospinal fibres. Both influence muscles of the extremities. Vestibulospinal and recticulospinal tracts that descend ventromedially in the spinal cord are concerned with posture and equilibrium through their influence on the trunk muscles.

126 Locomotion in its various forms (e.g. walking, running) consists of complex patterns of limb muscle activation accompanied by trunk and neck muscle contractions which are required to preserve balance.
   (a) The organisation of the CNS neurones which produce the principal patterns of limb muscle activity in locomotion are determined genetically and do not require to be learnt.
   (b) The principal pattern generator for locomotion lies in the cerebellum.
  *(c) Mildly noxious, simultaneous stimulation of both hind limbs in a high spinal (C1 section) cat elicits withdrawal (flexion) of both hind limbs. Crossed extension is suppressed.
  *(d) A decerebrated animal cannot walk because it lacks the source of initial drive to the spinal cord which comes from the cerebral cortex.
   (e) Afferent input from cutaneous receptors and proprioceptors can modify on-going locomotor activity.

*127 Studying disorders in the CNS caused by disease or physical damage can reveal much about the physiological function of different structures. This is especially so in a complex region such as the basal ganglia.
   (a) Disease of the basal ganglia may result in hypokinetic motor performance.
   (b) Parkinson's disease is a syndrome caused by degeneration of nigrostriatal neurones.
   (c) Tremor which is not present at rest but appears on commencing a movement is characteristic of Parkinson's disease.
   (d) Treatment with 5-hydroxytrytophan, the precursor of the neurotransmitter 5-hydroxytryptamine (5-HT), relieves Parkinson's symptoms.
   (e) Oral dopamine relieves Parkinson's symptoms.

126 (a) **True** Many quadrupeds walk and run at birth. Newborn human babies produce co-ordinated walking movements, if supported, but this facility is suppressed within weeks only to appear again at about one year of age. It is the co-ordination of balance and locomotion rather than the actual patterns of muscle activation that needs to be learnt by practice.

(b) **False** The pattern generators lie in the spinal cord. The cerebellum is important in the learning of the co-ordination necessary for efficient locomotion.

(c) **False** Such stimulation will elicit alternating, reciprocal flexion and extension in the limbs. The pattern of movement resembles stepping and is thought to be due to activation of the spinal locomotor pattern generators.

(d) **False** A decerebrated pigeon may walk incessantly. A decerebrated frog walks, jumps and will swim if placed in water. An area of the mesencephelon at the level of the superior colliculus is an important CNS element in the initiation of locomotion. This is intact in these decerebrated preparations.

(e) **True** Such reflex modification is essential to accommodate an uneven terrain or to react to obstacles. For example, a sudden increase in the depth of step alters proprioceptive discharge with a resulting prolongation of the motor discharges of the swing phase of the ipsilateral leg and the stance phase of the contralateral leg.

127 (a) **True** Parkinson's disease is an example. Degeneration of the basal ganglia can also result in hyperkinetic disorders. For example, damage to the caudate nucleus results in chorea. Chorea is characterized by rapid, intermittent dancing movements.

(b) **True** The substantia nigra and the striatum are both structures which form part of the basal ganglia.

(c) **False** A slow (4 Hz) tremor is present at rest in Parkinson sufferers. During a movement tremor is reduced.

(d) **False** The nigrostriatal neurones which have degenerated normally contain dopamine, not 5-HT.

(e) **False** L-DOPA, the metabolic precursor of dopamine, is used to alleviate Parkinson's symptoms.

128 Man maintains a constant body temperature despite large fluctuations in the
temperature of the environment.
  (a) The detectors of change in body core temperature are located in the skin.
  (b) The thermoreceptors of the skin are of one class in that they are homo-
      geneous in their response to cold and heat.
  (c) The skin is involved more with the detection of cold than with heat.
  (d) Loss of temperature sensitivity would indicate to the clinician damage to the
      dorsal columns of the spinal cord.
  (e) Heat loss through sweating is controlled by the parasympathetic division of
      the autonomic nervous system.

129 The following are statements about the autonomic nervous system:
  (a) The ratio of the number of preganglionic: postganglionic fibres is about 20:1.
  (b) The adrenal medulla secretes hormones with actions like those of the post-
      ganglionic nerves of the sympathetic nervous system.
  (c) The highest centre involved in integration of the autonomic nervous system
      is in the medulla oblongata.
  (d) Transmission velocity in postganglionic autonomic nerves is about the same
      as that in somatic motor nerves.
  (e) The effects of activity in parasympathetic nerves are more localised than
      those of sympathetic activity.

130 The following processes are brought about by activation of the parasympathetic
nerve fibres:
  (a) Defaecation.
  (b) Micturition.
  (c) Sweating.
  (d) Ejaculation of semen.
  (e) Dilation of the pupil.

131 The control of body weight is, despite the impression given by the human race,
a finely adjusted control system which is mediated via the central nervous
system.
  (a) Chemoreceptors are found in the stomach, liver and intestines which
      respond to levels of glucose, fats and amino acids.
  (b) Satiety, after ingestion of food, is partly the result of peripheral mecha-
      noreceptors responding to distension of the stomach.
  (c) Gastrectomy (removal of the stomach) substantially alters the amount of food
      ingested per day.
  (d) Destruction of part of the hypothalamus in rats can result in the animals
      ceasing to eat.
  (e) Electrical stimulation of a discrete area in the hypothalamus can cause rats
      not to eat.

128 (a) **False** Skin receptors are important for detecting changes in ambient temperature but body core temperature is sensed by neurones in the hypothalamus.

(b) **False** There are specific thermoreceptors which discharge optimally either to cold or hot skin temperatures.

(c) **True** Cold receptors outnumber warm receptors in the skin by as many as 10:1.

(d) **False** Temperature (and pain) afferents ascend the spinal cord in the contralateral spinothalamic tracts, not in the ipsilateral dorsal column. The latter conveys information from the other skin receptors and proprioceptors.

(e) **False** Sudomotor output is controlled by the sympathetic nervous system. So is the state of vasoconstriction in the skin — another important control of heat exchange.

129 (a) **False** There are always more postganglionic than preganglionic fibres. The ratio varies from 1 preganglionic to 1 or 2 postganglionic fibres in some parasympathetic ganglia to 1: 20 or 30 in some sympathetic ganglia.

(b) **True** Adrenaline and noradrenaline.

(c) **False** The hypothalamus is the most important higher integrating centre.

(d) **False** Much less. Postganglionic autonomic fibres are small ($0\cdot5$ to 1 μm diameter) and unmyelinated.

(e) **True** Work out why from the anatomy and mode of destruction of transmitters.

130 (a) **True**

(b) **True**

(c) **False** Sympathetic cholinergic fibres.

(d) **False** Sympathetic fibres.

(e) **False** Sympathetic fibres.

131 (a) **True** It has been proposed that their sensory discharges contribute to the feeling of hunger.

(b) **True** Inflating a balloon in the stomach will create a temporary state of satiety.

(c) **False** Smaller amounts of food are eaten but eating occurs more frequently so the total amount eaten stays fairly constant. Thus, gastric afferents are not the most important source of control signals for regulating food intake.

(d) **True** Aphagia (lack of eating) results if the lateral hypothalamus is lesioned. This area receives inhibitory afferent input from receptors responding e.g. to gastric distension.

(e) **True** Even if the rats have been starved, electrical stimulation of the medial hypothalamus causes aphagia.

132 The following are statements about vasomotor nerve fibres:
   (a) Sympathetic vasoconstrictor nerve fibres supply the smooth muscle in the walls of arterioles almost everywhere in the body.
   (b) The transmitter released by vasoconstrictor nerve fibres is noradrenaline.
   (c) There are also sympathetic vasodilator nerve fibres that supply arterioles in striated muscles.
   (d) The transmitter released by these vasodilator nerve fibres is adrenaline.
   (e) Vasodilator nerve fibres control sweating.

133 The urinary bladder and its internal sphincter are composed of smooth muscle. There is also an external sphincter composed of striated muscle.
   (a) The bladder is a reflex organ and contracts in response to stretch.
   (b) The parasympathetic nervous system causes contraction of the bladder.
   (c) Voluntary control of micturition involves direct interaction between the cerebral cortex and the spinal cord neurones which innervate the bladder and its sphincters.
   (d) During voluntary micturition there is inhibition of motoneurones innervating the external sphincter.
   (e) The bladder and internal sphincter receive inhibitory efferent control from the sympathetic nervous system which prevents micturition from occurring during sexual intercourse.

134 The following are statements concerning the nervous control of the cardiovascular system:
   (a) The peripheral baroreceptors of the carotid sinus and aortic arch regulate blood pressure via a spinal cord reflex.
   (b) The receptors in the carotid sinus and aortic arch respond with increased firing rate to a lowering of blood pressure.
   (c) A decrease in discharge rate of baroreceptor afferents results in an increase in heart rate and blood pressure.
   (d) Adjustments to the peripheral resistance in the control of blood pressure are directed to skeletal muscle rather than skin vasculature.
   (e) The sympathetic nervous system normally exerts a phasic modulation of vasomotor tone in skeletal muscle.

132 (a) **True**
    (b) **True**
    (c) **True**
    (d) **False** The transmitter is acetylcholine but adrenaline released from the adrenal medulla does act on β receptors to cause vasodilatation in the muscle vessels.
    (e) **False** There are sympathetic sudomotor nerve fibres that control sweating.

133 (a) **True** There is a component due to direct stretch activation of smooth muscle, and one due to a neural reflex.
    (b) **True** Preganglionic parasympathetic efferents lie in the sacral spinal cord. They receive visceral afferents from the bladder and both excitatory and inhibitory control from the brain.
    (c) **False** There is a pontine micturition centre which receives visceral afferent information from the bladder and can initiate involuntary emptying of a full bladder. This centre is under inhibitory control from the cerebral cortex.
    (d) **True** Relaxation of the striated muscle of the external sphincter is necessary for it to be opened. The smooth muscle of the bladder contracts.
    (e) **True** Closure of the urethra also ensures that semen does not enter the bladder during ejaculation.

134 (a) **False** Baroreceptor afferents travel in cranial nerves (glossopharyngeal and vagus) and their actions are mediated via the medulla. This does not, however, exclude the spinal cord from having a role in vasomotor control.
    (b) **False** The receptors lie in the walls of the vessels and respond to being stretched, i.e. to an increase in blood pressure, by increasing their discharge rate.
    (c) **True** The baroreceptors behave as stretch receptors. A reduction in their discharge rate, which would result from a fall in blood pressure, will increase vasomotor activity.
    (d) **True** Microelectrode recordings in intact man have shown that beat by beat adjustments are made to the peripheral resistance in skeletal muscle vasculature. Recordings from skin fascicles show mainly sudomotor (sweat) activity which is not related to the heart cycle.
    (e) **True** The vast skeletal muscle vascular bed is used as a variable resistance to correct moment by moment variations in blood pressure.

135 The following statements are concerned with the higher centres of nervous control of the circulation:
(a) Short-term control is exercised by the hypothalamus.
(b) Baroreceptor nerve impulses reduce the tonic output of medullary pressor neurones.
(c) Electrical stimulation of the rostral medulla elevates blood pressure.
(d) Assuming that respiration is maintained with a ventilator, low blood pressure will result from transection of the spinal cord at the first cervical vertebra.
(e) No significant cardiovascular changes can be elicited by electrical stimulation of the cerebral cortex.

136 This question concerns the peripheral and central nervous control of breathing.
(a) Neurones in the CNS which show a strong respiratory rhythm are found predominantly in the pons.
(b) Section of the vagal nerves would slow the rate of breathing.
(c) Specialized neurones in the medulla which act as chemoreceptors detect changes in the partial pressure of oxygen.
(d) A spinal cord transection at C6 would prevent breathing.
(e) The diaphragm is composed largely of smooth muscle innervated via the sympathetic nervous system.

137 This question is about the blood supply of the brain and the cerebrospinal fluid (CSF), which is found in the cerebral ventricles, the cisterns around the brain and in the subarachnoid space around the brain and spinal cord.
(a) The blood vessels in brain are as permeable as those in muscle to Na^+, K^+, Ca^{2+} and Mg^{2+} ions.
(b) Lipid soluble substances pass more rapidly into the brain from the cerebral vessels than do aqueous soluble substances.
(c) The local blood flow in brain regions is decreased when the brain tissue PCO_2 increases.
(d) CSF is an ultrafiltrate of plasma.
(e) When K^+, Ca^{2+} and Mg^{2+} ion concentrations change in the interstitial fluid outside the brain, their concentrations in the CSF alter to the same degree.

138 The most obvious circadian (daily) rhythm which humans, and many other animals, display is the waking–sleeping cycle.
(a) In sleep the neural activity of the brain is markedly reduced.
(b) Sleep is characterized by a clear alpha rhythm in the EEG.
(c) A person is more likely to report that he was dreaming if deliberately woken from rapid eye movement (REM) sleep than from non-REM sleep.
(d) Destruction of neurones containing 5-hydroxytryptamine (serotonin) in the raphe nucleus results in excessive sleeping.
(e) Destruction of neurones in the locus coeruleus results in a specific reduction in REM sleep.

135 (a) **False** The most important central nervous system sites of immediate cardiovascular control lie in the medulla oblongata.
   (b) **True** They also stimulate neurones in the caudal medullary depressor area.
   (c) **True** Stimulation of the caudal part of the medulla lowers blood pressure.
   (d) **True** Blood pressure falls as a result of separating tonically active medullary neurones from the sympathetic efferent nerves.
   (e) **False** Cortico-hypothalamic pathways to the medulla are implicated in behavioural and emotional responses of the cardiovascular system such as blushing and fainting.

136 (a) **False** They are found in the medulla. Some reticular neurones in the pons are involved in more specialized control of the airways but do not show a strong respiratory rhythm.
   (b) **True** The vagi contain afferents from pulmonary stretch receptors which inhibit inspiratory neurones in the medulla. Inspiration is prolonged so breathing is slowed.
   (c) **False** They are excited by an increase in partial pressure of carbon dioxide.
   (d) **False** Breathing with the diaphragm would continue. Only the intercostal muscles would be paralyzed since their motoneurones lie in the thoracic spinal cord.
   (e) **False** It is composed entirely of striated muscle.

137 (a) **False** There is a blood–brain barrier (BBB) to these ions (and many other substances). They do not diffuse freely from the cerebral vessels into the brain.
   (b) **True** There is no BBB to CO_2, O_2, alcohol, anaesthetics and other lipid soluble substances. They can pass rapidly across the epithelial cell membranes into the brain.
   (c) **False** A local increase in PCO_2 or, less effective, a decrease in PO_2 causes local dilatation of blood vessels.
   (d) **False** Many substances are actively secreted into and/or out of the CSF, and hence their concentrations are different from those in an ultrafiltrate of plasma.
   (e) **False** The CSF concentrations of K^+ and Ca^{2+} ions are lower than in interstitial fluid outside the brain, whilst that of Mg^{2+} ions is higher. The concentrations are maintained within narrower limits than in plasma and extracerebral interstitial fluid.

138 (a) **False** Neural activity is just as intense during sleep as during the awake state.
   (b) **False** Alpha rhythm (10 Hz) is typical of EEG recording from a relaxed awake person. On falling asleep the EEG changes progressively over many minutes from low amplitude, high frequency (10 Hz) waves to the higher amplitude, low frequency delta waves (1–3 Hz).
   (c) **True** REM sleep is associated with periods of high frequency, low amplitude waves in the EEG recording. It is not established, however, whether rapid eye movements are invariably a consequence of dreaming. If a subject is deprived of REM sleep by waking them during the REM sleep periods (or by giving sleeping pills), then they spend proportionally longer in REM sleep on subsequent nights.
   (d) **False** It results in insomnia. There is a reduction in both REM and slow wave sleep.
   (e) **True** These neurones contain noradrenaline.

Questions 139 and 140 concern the cerebral cortex:

139 (a) Activity in corticofugal neurones of the primary somatosensory receiving area of the cerebral cortex can inhibit synaptic transmission in the dorsal column nuclei.
   *(b) Loss of ability to recognize an object by feeling it in the left hand could be due to a lesion in the right parietal lobe.
    (c) The corpus callosum transfers information from one cortical hemisphere to a homologous region in the other hemisphere.
    (d) Electrical stimulation of the exposed occipital lobe in conscious humans can evoke specific memories.
   *(e) A lesion in the prefrontal cortex can alter the emotional significance of events.

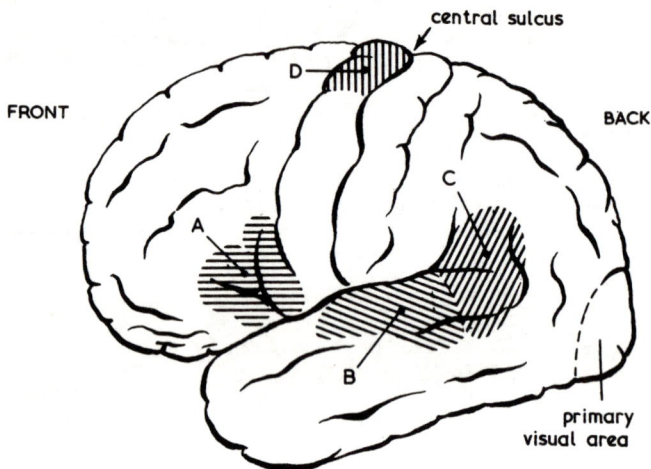

Lateral view of left hemisphere

*140 This question concerns speech and language. The hatched lines indicate areas where lesions have destroyed brain function.
    (a) Cerebellar lesions cannot affect speech.
    (b) A lesion in A, Broca's area, will affect comprehension of spoken language.
    (c) A lesion in D, In precentral cortex, will affect the control of voluntary movements of mouth and tongue.
    (d) A lesion in C, in the angular gyrus, may cause failure to understand written language.
    (e) The lips and tongue can be moved voluntarily (e.g. in eating), in spite of a lesion in A.

139 (a) **True** This is an example of the descending control of afferent input which is found in all sensory systems.

(b) **True** The condition is called astereognosia.

(c) **True** Learned information obtained using one hand, for example, is stored in both hemispheres, being transferred from the sensory and motor area contralateral to the hand, across the callosum to the other hemisphere. If the fibres of the corpus callosum are cut, an object felt by the right hand cannot be consciously matched with a similar one chosen only by feeling with the left hand.

(d) **False** Electrical stimulation of the *temporal* lobes has reproducibly elicited accounts of specific memories. Although this was established in epileptic patients undergoing surgery, memories could be elicited from apparently normal areas of cortex away from the epileptic focus.

(e) **True** Patients with prefrontal lesions may react angrily to the slightest provocation, and they are also said to be less able to concentrate on tasks than normal.

140 (a) **False** Lesions of the lateral and paravermal lobes may affect the coordination and accuracy of movements and hence speech may be slurred and slow.

(b) **False** A lesion in B, Wernicke's area, may produce this deficit. Broca's area A, is concerned with the motor control of speech. It is found in only one hemisphere, which for 90% of the population is on the left.

(c) **False** The leg and trunk are represented in this region of the precentral gyrus; the mouth and tongue areas are more lateral and inferior.

(d) **True** Damage here can produce dyslexia, even when the temporal lobe portion of the interpretive area B is intact. Note that area C lies between the visual cortex and Broca's area.

(e) **True**